AF887402

DIGESTION

International Journal of Gastroenterology

Founded as 'Archiv für Verdauungskrankheiten' 1895 by I. Boas
Continued as 'Gastroenterologia' 1939–1967
Former Editors: P. Morawitz (1934–1936), R. Staehelin (1937–1943), A. Hurst (1940–1945), W. Löffler (1943–1961), Th.C. Hunt (1947–1967), N. Henning (1953–1962), B. Ihre (1953–1967), H. Bartelheimer (1963–1967), M. Demole (1963–1971), H. Kapp (1968–1970), R. Lambert (1972–1978)

KARGER

Printed in Switzerland
on acid-free paper by
Thür AG Offsetdruck, Pratteln

Appears monthly:
3 volumes per year (12 issues)

Submission: 'Digestion' invites contributions concerning all aspects of the digestive tract and liver. Only original papers written in English are considered and should be sent to:

Prof. W. Creutzfeldt
Medizinische Klinik der Universität
Robert-Koch-Strasse 40
D-W-3400 Göttingen (FRG)

Manuscripts should be submitted in triplicate (with three sets of illustrations of which one is an original), typewritten double-spaced on one side of the paper, with a wide margin.

Conditions: All manuscripts are subject to editorial review. Manuscripts are received with the explicit understanding that they are not under simultaneous consideration by any other publication. Submission of an article for publication implies the transfer of the copyright from the author to the publisher upon acceptance. Accepted papers become the permanent property of 'Digestion' and may not be reproduced by any means, in whole or in part, without the written consent of the publisher. It is the author's responsibility to obtain permission to reproduce illustrations, tables, etc. from other publications.

Special Requirements:
1. Investigations Involving Human Subjects: 'Digestion' requires that investigations performed on human subjects have the prior approval of the institutional committee on human experimentation. Authors are required to submit a signed statement as to the date and details of the appropriate review. In countries where such mechanisms for approval do not exist, authors are required to submit a signed declaration that the research was carried out in accordance with the Helsinki Declaration.
A similar assurance is required to document that animal experimentation was performed under appropriate circumstances.
2. Statement for Authors Submitting Original Research or Case Reports: 'The enclosed manuscript . . .(Title). . . has been approved by me as well as by the responsible authorities at the institute where the work has been carried out.
I certify that none of the material in this manuscript has been published previously in any form and that none of this material is currently under consideration for publication elsewhere. This includes symposia and proceedings of meetings and preliminary publications of any kind except an abstract of 400 words or less.'
(To be signed by *all* authors)
3. The undersigned also confirm that they qualify for (co-) authorship according to the 'Uniform Requirements for Manuscripts' published in the guidelines of the 'International Committee of Medical Journal Editors' in 1988 as follows: Authorship credit should be based only on substantial contributions to (a) conception and design, or analysis and interpretation of data; (b) drafting the article or revising it critically for important intellectual content; and on (c) final approval of the version to be published. Conditions (a), (b), and (c) must all be met. Participation solely in the acquisition of funding or the collection of data does not justify authorship. General supervision of the research group is also not sufficient for authorship. Any part of an article critical to its main conclusions must be the responsibility of at least one author. Persons who have contributed intellectually to the article but whose contributions do not justify authorship may be named and their particular contribution described in the section 'Acknowledgements'. Such persons must give their permission to be named.

Arrangement:
Title page: The first page of each paper should indicate the title (main title underlined), the authors' names, and the institute where the work was conducted. A short title for use as running head as well as the full address of the author to whom correspondence should be sent are also required.
Full address: The exact postal address complete with postal code of the senior author must be given at the bottom of the page. Please also supply Phone and Fax numbers.
Key words: For indexing purposes, a list of 3–10 key words in English is essential.
Abstract: Each paper needs an abstract of up to 10 lines.
Small type: Paragraphs which can or must be set in smaller type (case histories, test methods, etc.) should be indicated with a 'p' (petit) in the margin on the left-hand side.
Footnotes: Avoid footnotes. When essential, they are numbered consecutively and typed at the foot of the appropriate page.
Tables and illustrations: Tables and illustrations (both numbered in Arabic numerals) should be prepared on separate sheets. Tables require a heading and figures a legend, also prepared on a separate sheet. For the reproduction of illustrations, only good drawings and original photographs can be accepted; negatives or photocopies cannot be used. When possible, group several illustrations on one block for reproduction (max. size 144 × 196 mm) or provide crop marks. On the back of each illustration, indicate its number, the author's name, and 'top'. Color illustrations are reproduced at the author's expense.
References: In the text identify references by Arabic numerals [in square brackets]. Material submitted for publication but not yet accepted should be noted as 'unpublished data' and not be included in the reference list. The list of references should include only those publications which are cited in the text. Do not alphabetize; number references in the order in which they are first mentioned in the text. The surnames of the authors followed by initials should be given. There should be no punctuation other than a comma to separate the authors. Cite all authors, 'et al' is not sufficient. Abbreviate journal names according to the Index Medicus system. (Also see International Committee of Medical Journal Editors: Uniform requirements for manuscripts submitted to biomedical journals. Br Med J 1988;296:401–405.)
Examples
(a) Papers published in periodicals: Kauffman HF, van der Heide S, Beaumont F, Blok H, de Vries K: Class-spe-

Instructions to Authors (Continuation)

cific antibody determination against *Aspergillus fumigatus* by means of the enzyme-linked immunosorbent assay. III. Comparative study: IgG, IgA, IgM ELISA titers, precipitating antibodies and IgE binding after fractionation of the antigen. Int Arch Allergy Appl Immunol 1986;80:300–306.
(b) Monographs: Matthews DE, Farewell VT: Using and Understanding Medical Statistics. Basel, Karger, 1985.
(c) Edited books: Hardy WD Jr, Essex M: FeLV-induced feline acquired immune deficiency syndrome: A model for human AIDS; in Klein E (ed): Acquired Immunodeficiency Syndrome. Prog Allergy. Basel, Karger, 1986, vol 37, pp 353–376.

Further details: Authors may consult our leaflet 'Rules for the Preparation of Manuscripts', available on request from the Publisher, S. Karger AG, P.O. Box, CH – 4009 Basel (Switzerland).

Page charges: There is no page charge for papers of 5 or fewer printed pages (including tables, illustrations and references). Each additional complete or partial page is charged to the author at SFr. 295.– / US $ 195.00. The allotted size of a paper is equal to approx. 10 manuscript pages (including tables, illustrations and references). The inability to pay the additional page charge will not prejudice publication of the paper. This will be decided by the Editor.

Galley proofs: Unless indicated otherwise, galley proofs are sent to the first-named author and should be returned with the least possible delay. Alterations made in galley proofs, other than the correction of printer's errors, are charged to the author. No page proofs are supplied.

Reprints: Order forms and a price list are sent with the galley proofs. Orders submitted after the issue is printed are subject to considerably higher prices. Allow five weeks from date of publication for delivery of reprints.

General Information

DIGESTION

Publication data: 'Digestion' is published 12 times annually. Volumes 51–53 with 4 issues each appear in 1992.
Subscription rates: Subscriptions run for a full calendar year. Prices are given per volume, surface postage included.
Personal subscription:
SFr. 179.20, US$ 119.70.
(Must be in the name of, billed to, and paid by an individual. Order must be marked 'personal subscription'.)
Institutional subscription:
(Regular rate.) SFr. 256.–, US$ 171.00.
Airmail postage: SFr. 25.–, US$ 20.00.
(Extra per volume.)
Microform subscription:
SFr. 128.–, US$ 85.50.
(Available exclusively to subscribers of the paper edition.)
Single issues and back volumes: Information on availability and prices can be obtained through the Publisher.
Advertising: Correspondence and rate requests should be addressed to the Publisher.

Subscription orders: Orders can be placed at agencies, bookstores, directly with the Publisher S. Karger AG, P.O. Box, CH – 4009 Basel (Switzerland)
or with any of the following distributors:
Deutschland,
S. Karger GmbH, Lörracher Strasse 16A
Postfach, D-W-7800 Freiburg
France: Librairie Luginbühl
36, bd de Latour-Maubourg, F-75007 Paris
Great Britain, Ireland:
S. Karger Publishers
11 Ravenscroft Road
Beckenham, Kent BR3 4TP
USA: S. Karger Publishers, Inc.
26 West Avon Road, P.O. Box 529
Farmington, CT 06085
India, Bangladesh, Sri Lanka:
S. Karger India Subscription Agency
B – 5/132 Safdarjung Enclave
New Delhi-110029
Japan: Katakura Libri, Inc.
36-9 Hongo 3-chome, Bunkyo-ku
Tokyo 113
Change of address: Both old and new address should be stated and sent to the subscription source.

Bibliographic indices: This journal is regularly listed in bibliographic services, including *Current Contents®*.
Copying: This journal has been registered with the Copyright Clearance Center (CCC), as indicated by the code appearing on the first page of each article. For readers in the US, this code signals consent for copying of articles for personal or internal use, or for the personal or internal use of specific clients, provided that the stated fee is paid per copy directly to CCC, 21 Congress Street, Salem, MA 01970 (USA). A copy of the first page of the article must accompany payment. Consent does not extend to copying for general distribution, for promotion, for creating new works, or for resale. In these cases, specific written permission must be obtained from the copyright owner, S. Karger AG, P.O. Box, CH – 4009 Basel (Switzerland).

KARGER

Theilheimer's Synthetic Methods of Organic Chemistry
(ISSN 0253–200X), **Vol. 46**
Series Editor: A.F. Finch, London

Its universal coverage and brilliant organization make this book an indispensable reference tool for every need... and a delight to browse.

Theilheimer's
Synthetic Methods
of Organic Chemistry
10th series

46/1992 Yearbook

KARGER

Theilheimer – your portable data base!

Also available:

Theilheimer's Synthetic Methods of Organic Chemistry: A Guide for Users

Compiled by Alan F. Finch and Paul R. Mitchell

Write for your free copy!

'The appearance of an annual volume of Theilheimer is an event always eagerly anticipated by organic chemists, because of its comprehensiveness and highly developed organization. It is one of the major entries into the recent literature, and there can be few chemists who do not find a lot of information that they would otherwise have missed . . . The production of this book is excellent . . .' Journal of the American Chemical Society

Theilheimer's Synthetic Methods of Organic Chemistry, Vol. 46
Yearbook 1992
Editor: Finch, A.F. (London)
XXII + 506 p., hard cover, 1992
SFr. 798.– / DM 956.– / £ 347.00 / US $ 638.50
Prices subject to change
US $ price for USA, £ price for UK, DM price for FRG only
ISBN 3-8055-5509-1

KARGER

KI 92606

EPC
European Pancreatic Club

Abstracts

XXIVth Meeting
Ulm, Germany
October 11–14, 1992

Editors
H.G. Beger, Ulm, Germany
M. Büchler, Ulm, Germany

KARGER

S. Karger
Medical and Scientific Publishers
Basel · Freiburg · Paris · London
New York · New Delhi · Bangkok
Singapore · Tokyo · Sydney

Drug Dosage
The authors and the publisher have exerted every effort to ensure that drug selection and dosage set forth in this text are in accord with current recommendations and practice at the time of publication. However, in view of ongoing research, changes in government regulations, and the constant flow of information relating to drug therapy and drug reactions, the reader is urged to check the package insert for each drug for any change in indications and dosage and for added warnings and precautions. This is particularly important when the recommended agent is a new and/or infrequently employed drug.

Printed in Switzerland on acid-free paper by
Thür AG Offsetdruck, Pratteln
ISBN 3-8055-5693-4

Abstracts

Digestion 1992;52:67–137

1 Mortality in Pancreatitis

R. Andersson[a], I. Ihse[a], Å. Andrén-Sandberg[a], P. Krantz[b], P. Löwenhielm[b], I. Plantin
Departments of [a]Surgery and [b]Forensic Medicine, University of Lund, Sweden

Acute pancreatitis (AP) is still associated with a substantial morbidity and mortality, though mortality rates reported vary widely. Little is known about death outside hospital due to AP.

Patients. 14 out of 519 patients (2.7%; 12 men, 2 women; mean age 62 (37–83) years) admitted to the Department of Surgery, Lund, due to AP died during the period 1981–1990. In 36 other cases (30 men, 6 women; mean age 56 (29–83) years) death outside hospital was caused by AP 1984–1990.

Results. Among patients with AP dying following admission to hospital, pancreatitis was induced by alcohol (6), biliary stones (5), trauma (1), postoperative (1) or unknown cause (1). Only one patient had had a previous attack of pancreatitis. Seven out of 14 patients were operated on, the first operation being done after 8 (4–15) days; in mean 2 (1–4) operations were performed per patient. Twelve out of 14 developed multiple system organ failure (MSOF), including respiratory (10), renal (12), cardiac (9) and hepatic (6) failure, gastrointestinal bleeding (2) and coagulopathia (3). Death was considered due to MSOF in 12 and circulatory failure in 2 and occurred after 21 (2–73) days. In the group with AP dying outside hospital, alcohol-induced pancreatitis dominated (27/36). All had necrotizing or hemorrhagic pancreatitis (2 patients had 1000 and 2000 ml of blood, respectively, in the peritoneal cavity at autopsy). Fifteen out of 27 with alcohol-induced AP had detectable levels of ethanol in cardiac blood.

Conclusion. The mortality noted in AP is mainly due to MSOF. Repeated laparotomy is recommended in patients with severe AP and signs of necrosis/abscess, especially if seen together with developing organ failure. Mortality in AP outside hospital is mainly caused by alcohol in subjects with known alcohol abuse.

2 Recurrence after Pancreatectomy for Pancreatic Cancer

Å. Andrén-Sandberg, J. Westerdahl, I. Ihse
Department of Surgery, University of Lund, Sweden

The results of radical surgery for pancreatic cancer depend, like all other cancer surgery, on the technique used and the biology of the disease. We have analysed the site and time of recurrence after pancreatectomy for exocrine pancreatic cancer in 74 patients who died more than 6 months postoperatively. All patients had recurrent disease, 64 had local recurrence in the pancreatic bed and 68 had liver metastases. Local recurrence without liver metastases was found in 6 patients, and 10 had liver metastases but no local recurrence. The time from operation to clinically evident recurrence and the postoperative survival time, respectively, were significantly longer for patients with local recurrence only. Although not statistically significant there was a tendency (5 out of 8) for smaller, well differentiated tumors without spread outside the pancreas to give local recurrences without liver metastases. We conclude that, in retrospect, the surgical procedures used were inappropriate and insufficient. To cure these patients a more radical operation and/or effective adjuvant treatment is needed.

3 Management of Cysts and Pseudocysts Complicating Acute and Chronic Pancreatitis: A Retrospective Study of 188 Patients

M. Barthet, J.P. Bernard, C. Bastid, J. Sahel
Service d'hépatogastroentérologie, Hôpital Sainte-Marguerite, Marseille, France

Forty-five patients with 55 pancreatic pseudocysts (PC) complicating acute pancreatitis (AP) and 143 patients with 170 cysts (C) or pseudocysts (PC) complicating a

0012–2823/92/
0522–0067 $2.75/0

chronic pancreatitis (CP) were followed between 1980 and 1990. 8% of C and PC complicating CP disappeared spontaneously: they were little cysts (average 28 mm), communicating with the ductal system in 60% of the cases. Resolution of the pseudocysts was observed in 5 patients among 45 with AP. Percutaneous punctures were performed in 47 patients presenting CP and in 26 patients presenting AP, with a low morbidity (5%). It was an effective treatment in 33% of C and PC complicacy CP and in 42% of PC complicating AP. Percutaneous drainage was performed in 9 patients with CP with a morbidity of 1/9, a success rate of 5/9, and in 9 patients presenting AP with a morbidity rate of 3/9 and a success rate of 7/9. Seventy-eight endoscopic cystoenterostomies were performed in 71 patients with CP. Twelve (15.3%) had complications and one died (1.3%). The success rate was 72.2%. Twelve patients with AP underwent endoscopic cystoenterostomy. Three complications (25%) were observed and 1 patient died. Seven out of 12 patients did not require further treatment. Finally surgery was needed in 63/143 (44.05%) patients with CP and in 13/45 (28.8%) patients with AP, mostly internal drainage. In AP, 1 major complication occurred and 1 patient died, while in CP 8 complications occurred (12.6%) and 1 patient died (9.3%). Percutaneous drainage and endoscopic cystoenterostomy appear as effective treatments for C and PC complicating AP or CP. Surgery should be considered in case of failure or technical impossibility of percutaneous or endoscopic approaches.

4 Antibiotics Penetration in Human Necrotizing Pancreatitis (NP)

C. Bassi, P. Pederzoli, S. Vesentini, G. Cavallini, P. Bovo, D. Lombardi, H. Abbas, F. Nifosì, A. Benini, E.M. Bertazzoni
Surgical, Medical and Pharmacological Departments, Policlinico, University of Verona, Italy

The choice of antibiotics in preventing infected necrosis during NP should be based on their antimicrobial activity, penetration and therapeutic concentrations in necrotic pancreatic tissue (NPT).

The aim of the study was to determinate whether or not antibiotics excreted by the normal pancreas are also excreted in human NP, reaching the tissue sites of the infection. Nine patients suffering from acute necrotizing pancreatitis were treated with imipenem (Im, 0.5 g), mezlocillin (M, 2 g), gentamicin (G, 0.08 g), amikacin (A, 0.5 g), pefloxacin (P, 0.4 g) and metronidazole (Mz, 0.5 g). Serum and NPT samples were collected at the same time 2 h after in vitro drug administration, by CT-guided needle aspiration, intraoperatively and from surgical drains. Drug concentrations were determined by HPLC and microbiological assay.

All antimicrobial drugs reached the NPT, but with varying degrees of penetration, this being low for aminoglycosides (13%) and high in the case of P (89%) and Mz (99%). The concentrations of P (2.05 μg/ml) and Mz (6.3 μg/ml) in NPT were distinctly higher than the MICs for the organisms most commonly isolated in this disease; the tissue concentrations of Im (3.35 μg/ml) and M (8.0 and 15.0 μg/ml) did not always exceed the MICs, whereas the aminoglycoside NPT concentrations (0.5 μg/ml) were inadequate. Repeated administrations seem to enhance drug penetration. These findings represent a further step forward in our understanding of the therapeutic guidelines in NP.

5 Role of Imipenem (I) in Preventing Infected Necrosis (IN) during Acute Pancreatitis (NP). Results of the Italian Multicenter Study

C. Bassi, V. Di Carlo, A. Zerbi, V. Galloro, G. Uomo, G. Fontana, R. Pezzilli, P. Mascagni, G.P. Marzoli, F. Martin, S. Vesentini, G. Cavallini, A. Campedelli, P. Pederzoli
The Italian Multicenter Study Group (Verona, Bologna, Bolzano, Milano, Roma, Napoli), Italy

Clinical studies of effective antibiotics for prophylaxis on IN do not currently exist. Recent evidence of pancreatic penetration at therapeutic MICs of several antimicrobial drugs active against the usual flora found in IN has prompted us to carry out a study on I prophylaxis in NP. In 6 Italian centers, 74 consecutive patients (44 men, 30 women; mean age 52; range 20–84) with CT scans demonstrating NP (mean Ranson score 3.7) within 72 h from onset were randomly assigned to 2 groups receiving either non-antibiotic treatment (NA) or prophylactic I (I) 0.5 g e.v. every 8 h for 2 weeks. Pancreatic sepsis was always detected by means of cultures (CT or US guided needle aspiration). The etiology was biliary in 37 cases, alcoholic in 24 and due to other causes in 13.

The overall incidence of IN was 15 patients (20.3%), 10 (30.3%) on NA vs. 5 (12.2%) on I ($p<0.01$ on the U test). No IN developed in I group when the necrosis was less than 50% of the pancreatic volume, whereas in NA group small amounts of necrosis (<30%) did not prevent the sepsis. Also the incidence of nonpancreatic sepsis was statistically different in favour of the I group, while multiorgan failure, need of surgery and mortality rate were superimposable.

In conclusion, the use of I significantly reduces the incidence of IN in NP.

6 Altered C Peptide Release after Glucagon Stimulation in Pancreatic Cancer

D. Basso[b], *P. Fogar*[a], *T. Meggiato*[a], *G. Del Favero*[a], *G. Briani*[a], *C. Ferrara*[a], *R. Naccarato*[a]
[a]Cattedra di Gastroenterologia et [b]Istituto di Medicina di Laboratorio, Università degli Studi di Padova, Italy

Hyperglycemia may be one of the first symptoms of pancreatic cancer (PC), but it is not clear, which alterations develop in the islet cells in PC. The aim of the present study was to evaluate the behaviour of serum C peptide (C-p) after i.v. glucagon stimulation in PC patients in comparison to control and patients with type I diabetes mellitus. We studied 18 subjects: 7 were control subjects (CS), 4 were affected by PC and 7 by type I diabetes mellitus (DM). Glucagon (1 mg) was bolus infused in fasting subjects; blood samples were taken before the above injection (time A) and after 6 min (time B). In 3 out of the 4 PC patients an other test stimulus was performed. After fasting a standard meal was given and blood samples were taken before the meal and after 10 and 30 min. C-p was determined in all serum samples by means of a RIA method. Basal values of C-p were significantly lower ($p<0.001$) in DM (0.13. ± 0.1 µg/l, $\bar{x} \pm SD$) as compared to CS (1.69 ± 0.3) and PC (1.73 ± 0.6). No significant difference was detected between CS and PC. After glucagon infusion C-p significantly increased in CS ($t=7.30$, $p<0.001$) but not in PC ($t=0.31$, p: ns) and DM ($t=1.53$, p:ns). The ratios between C-p values at time B and time A were significantly lower in PC and DM as compared to CS ($F=14.11$, $p<0.001$). No difference was observed between PC and DM. No variation in C-p serum levels was observed in the 3 patients with PC subjected to the meal test.

Conclusions. (1) In PC patients basal C-p varied in the normal range; (2) no variation in C-p serum levels was found in PC after glucagon stimulation, similarly to what happens in DM; (3) a similar behaviour was found after a meal stimulus. An altered β-cell answer to different stimuli in PC may be, therefore, suggested.

7 Mechanisms of Intraduodenal Fat and Amino Acid-Stimulated Exocrine Pancreatic Secretion in Man

C. Beglinger[a], *S. Mossi*[a], *P. Hildebrand*[a], *K. Gyr*[b]
[a]Division of Gastroenterology, University Hospital, Basel, and [b]Department of Internal Medicine, Kantonsspital, Liestal, Switzerland

Our studies were designed to analyze which mechanisms mediate pancreatic enzyme and bicarbonate secretions in response to intraduodenal (ID) sodium oleate and tryptophan (Trp) stimulation in man. CCK-A receptor blockade (loxiglumide = Lox) and atropine (Atr) were used as specific tools to test these patterns.

Methods. Twelve healthy male volunteers were enrolled. In part I ($n=6$), 3 experiments were performed in random order. After a 60-min basal period, increasing loads of sodium oleate (0.5, 2, 8 mmol/h) were sequentially perfused ID, each load for 45 min. Background infusions of saline (control), Lox (10 mg/kg/h) or Atr (5 µg/kg/h) were given on the 3 study days. In part II ($n=6$), increasing loads of Trp were perfused ID amounting to 1.1, 3, and 10 mmol/h with concomitant infusions of either saline, Lox or Atr, given on 3 different days. Exocrine pancreatic secretion was quantified with a multilumen duodenal tube and marker (PEG 4000) perfusion technique. CCK and secretin plasma levels were determined by RIA.

Results. Data are cumulative responses (amylase in KIU, bicarbonate in mmol/135 min to all 3 loads of stimulant, mean ± SEM). ID perfusion of oleate markedly ($p<0.05$) stimulated CCK (from 3.9 ± 0.6 to max. 12.5 ± 0.5 pg/ml) and secretin (from 0.9 ± 0.1 to 2.1 ± 0.3 pg/ml, $p<0.05$). Atr did not alter these responses, whereas Lox enhanced the CCK response 2-fold ($p<0.01$). ID Trp did not stimulate secretin, but induced a significant ($p<0.05$) CCK response (from 3.6 ± 0.4 to 8.5 ± 0.7 pg/ml). Lox again augmented this response 2-fold ($p<0.01$), whereas Atr had no effect.

	Oleate			Trp		
	saline	Lox	Atr	saline	Lox	Atr
Amylase	39 ± 6	15 ± 4*	6 ± 3*	27 ± 6	10 ± 3*	5 ± 2
Bicarb	25 ± 3	6 ± 2*	9 ± 3*	15 ± 3	5 ± 1*	8 ± 2*

* Indicates $p<0.05$ vs. control (ANOVA).

Conclusions. (1) CCK mediates much of the bicarbonate potentiation between secretin and intestinal oleate or Trp; (2) pancreatic enzyme secretion after ID oleate or Trp stimulation is predominantly under cholinergic control with endogenous CCK modifying the response.

8 The Role of C1-Esterase Inhibitor (C1) in Experimental Acute Pancreatitis (AP)

L. Benetti[a], *C. Bassi*[a], *S. Vesentini*[a], *A. Bonora*[a], *G. Zamboni*[b], *R. Girelli*[a], *M. Falconi*[a], *H. Abbas*[a], *A. Campedelli*[c], *P. Pederzoli*[a]
Departments of [a]Surgery, [b]Pathology and [c]Statistics, University of Verona, Italy

The regulatory protein C1 functions as an inhibitor of several proteases and interferes with the amplification of cascade systems, such as complement, kinins, coagulation and fibrinolysis. We planned the present study to verify the effects of C1 in experimental AP, induced by the technique of Aho in 180 Inbred Sprague-Dawley rats (R). Ninety R received no treatment (control group) and were sacrificed at 72 h (45 R) and at 7 days (45 R). The remaining 90 R were divided in 3 groups and treated respectively with a 100 UI C1 injection 15 min before the induction of AP (30 R) and an additional injection of 100 UI after 24 h (30 R) and after 24 and 48 h (30 R). In each group 15 R were scheduled for sacrifice at 72 h and 15 at 7 days. The pancreas of the surviving R was removed and blindly examined by the same pathologist to evaluate the following lesions: oedema, cellular necrosis, massive necrosis, cellular degeneration and regeneration indices. The mortality rate showed a highly significant ($p<0.005$) difference between treated and control R, regardless of C1 dosage. The histological signs of phlogosis, cellular damage and regeneration were not significantly affected by treatment, whereas the massive pancreatic necroses were positively reduced by treatment, with a highly significant ($p<0.005$) difference in treated vs. control group.

In conclusion, C1 seems to have a prophylactic effect against massive pancreatic necrosis in experimental AP. Since a vascular protective mechanism may be implied, we are planning further investigations to verify these data.

9 Nasopancreatic Drainage for Chronic Calcifying Pancreatitis (CCP)

Z. Berger, L. Topa, T. Takács, Å. Pap
2nd Department of Medicine, St. Imre Hospital, Budapest, and 1st Department of Medicine, University of Szeged, Hungary

In CCP, progression is promoted and regeneration is inhibited by obstruction due to intraductal protein plugs and stones. Insufficient outlet of pancreatic juice has a significant role in provoking hypertension and pain as well as pancreatic insufficiency. Endoscopic nasopancreatic drainage is technically simple, dissolution of stones and dilatation of strictures can be achieved in some days, the procedure can be repeated if necessary. We performed nasopancreatic drainage in 6 CCP patients (4 males, 2 females). Pancreatic duct was perfused with a mixture of isotonic citrate and saline at 3 ml/min for 4 days. Stone-free state was achieved in all cases. Pancreatic pain disappeared and 4 patients remained free of pain for the follow-up period (1–12 months), and a repeated treatment assured a long-term effect in other 2 cases. Pancreatic exocrine function was checked by Lundh test in all patients except one: an increase of 200 and 360% was observed in enzyme output of the 2 female patients, while the pancreatic function did not change in 2 males and slightly improved (50%) in another. In conclusion: Nasopancreatic drainage can be tried as first procedure to relieve pancreatic pain and obstruction due to strictures and intraductal stones.

10 Quantification of Human Lithostathine in Pancreatic Juice by High Performance Liquid Chromatography (HPLC)

J.P. Bernard, M. De Reggi, M. Barthet, J. Sahel, H. Sarles, J.C. Dagorn
INSERM U315, Marseille, France

Lithostathine is assumed to control $CaCo_3$ crystal growth in pancreatic juice, and a reduced lithostathine concentration could be one of the factors leading to chronic calcifying pancreatitis (CCP). However, there have been discrepancies in lithostathine concentration determined using immunological methods. These discrepancies could be due to differences in antibody affinity for the different forms of lithostathine in juice. In an attempt to overcome these difficulties inherent to immunological techniques, we used an HPLC method to quantify the protein.

Pure pancreatic juice samples were collected under secretin stimulation by retrograde cannulation of the pancreatic duct in 27 patients, namely CCP ($n=6$), CCP with radiolucent stones (CCP/RS; $n=2$), alcoholics without pancreatitis (A1; $n=7$), pancreatic diseases other than CCP (P ctrl; $n=3$) and controls without pancreatic or biliary diseases (Ctrl; $n=9$). Lithostathine was separated using a HPLC TSK-phenyl column. Proteins were quantified by UV absorbance at 280 nm.

Results. Mean lithostathine concentrations were expressed as micrograms of lithostathine per mg of total proteins: CCP, 115.0; CCP/RS, 295.0; A1, 109.2; P ctrl, 266.6; Ctrl, 272.2.

T test showed that lithostathine concentration was significantly reduced in CCP and alcoholic patients, as compared to controls (P ctrl+Ctrl) ($p=0.0001$). We underline that the 2 CCP/RS patients had a lithostathine concentration comparable to controls.

Conclusion. HPLC separation is a suitable method for the quantification of proteins in pancreatic juice, and the data confirm the decreased lithostathine concentration in CCP and alcoholic patients.

11 Endoscopic Treatment by Stenting of Dorsal Duct in 13 Cases of Pancreas Divisum

G. Bordes, J. Sahel
Service d'hépato-gastro-entérologie, Hôpital Sainte-Marguerite, Marseille, France

It is speculated that functional obstruction of the minor papilla may be the cause of recurrent acute pancreatitis associated with pancreas divisum (PD). We studied the role of endoscopic treatment by stent placement across the minor papilla.

Patients and Methods. Thirteen PD (10 men, 3 women; age: 31–73 years, mean 53±14). Nine had acute recurrent pancreatitis (1–4/year, mean 1.6); 4 had pancreatic-like epigastric pain. A minor papilla stricture with duct dilatation was observed in 3 cases. Other causes of acute pancreatitis and evidence of chronic pancreatitis were excluded before treatment. Following minor papilla cannulation and duct opacification, a straight prosthesis (5–10 F) with lateral holes and proximal and distal barbs has been inserted, over a wire guide, in the dorsal duct. A minor papilla sphincterotomy or resection occurred in 7 cases, dilatation in 3 cases. Each prosthesis stayed in place about 3.5 months, the total drainage time was 5 months per patient. Twenty-three stenting procedures were carried out in those 13 patients.

Results. Nine patients (69%) had no further attacks of pancreatitis or pain following stent placement during a follow-up period of 17 months (4–58 months). In 2 cases, pain improved but did not disappear. Best results were obtained in patients presenting with acute recurrent pancreatitis, compared with the epigastric pain group. Papilla stricture and duct dilatation always disappeared after stenting. No serious complication occurred. The stent migrated into the intestine in 3 patients and into the body of the pancreas in 1 patient. Clogged stents were found at the time of exchange twice.

Conclusion. Stenting dorsal duct is technically possible, associated with minimal complications, probably less than after endoscopic papillotomy of the minor ampulla. Good results obtained with such a treatment seem to confirm that obstruction is the cause of acute pancreatitis and pain observed in PD.

12 Endoscopic Treatment of Chronic Pancreatitis by Stenting of Wirsung Duct

G. Bordes, J. Sahel
Service d'hépato-gastro-entérologie, Hôpital Sainte-Marguerite, Marseille, France

Endoscopic stenting or endoprosthesis placement has been purposed in the treatment of chronic pancreatitis (CP) to achieve decompression in pancreatic duct.

Patients and Methods. The results of such treatment have been analyzed in 15 patients (11 men, 4 women; age: 15–63 years, mean 40±13): 11 had recurrent acute pain, 4 had continuous epigastric pain. A cephalic duct stricture was present in 6 cases, pancreatic stones in 8 cases, cysts in 6. After retrograde pancreatography (9 times after sphincterotomy), a straight prosthesis (7–10 F) with lateral holes and proximal and distal barbs has been inserted, over a wire guide, in the main pancreatic duct. The mean time of drainage was 5 months, each prosthesis was changed every 3.5 months. Twenty-two stenting procedures were carried out in those 15 patients.

Results. During a mean follow-up of 13 months (2 days to 36 months), 9 patients (60%) showed no recurrence of pain. In 3 cases, the symptoms persisted or aggravation lead to a surgical treatment. Best results were obtained in patients presenting with a duct stricture. When pancreatic stones were present, stenting results were improved by total extraction. The endocrine or exocrine pancreatic function was poorly changed by stenting. No serious complications resulting from treatment were observed: severe epigastric pain occurred in 2 cases, cholecystitis treated by transcutaneous punction in 1 case. Stent migration in the intestine was seen 3 times, and clogging of the prostheses in 4 cases.

Conclusion. Endoscopic stenting of the main pancreatic duct may be proposed in the management of CP, as an alternative to surgery. This treatment must be associated with other endoscopic procedures (sphincterotomy, stone extraction) and is particularly useful in cases of pancreatic duct stricture.

13 Postprandial Gastric pH Is Abnormally Low in Patients with Exocrine Pancreatic Insufficiency

P. Bovo[a], C. Cataudella[a], B. Vaona[a], G. Talamini[a], L. Frulloni[a], V. DiFrancesco[a], M.P. Brunori[a], M. Filippini[a], C. Bassi[b], M. Marcori[a], G. Cavallini[a]
[a]Clinica Medica and [b]Clinica Chirurgica, University of Verona, Italy

Increases in basal and pentagastrin-stimulated acid output have been reported in the course of chronic alcoholic pancreatitis (CAP), though it is still debatable whether or not any correlation exists between gastric secretory function and exocrine pancreatic insufficiency. The aim of this study was to assess the circadian variations in intragastric pH in 30 hospital patients with CAP (mean age 47±12.3 years) and in 14 controls (C) (mean age 45.4±9.8). pH-metry was performed using a single-crystal antimony electrode placed in the gastric corpus (fluoroscopic control) and connected up to a recorder (MKII Digitrapper, Synectis). The evaluation parameters considered, expressed as median (m) and interquartile range were: total period, post-prandial periods, interdigestive and nocturnal phases. The CAP patients were subdivided into 2 groups on the basis of severity of exocrine pancreatic insufficiency (S-Cae test: lipase output at 60–90 min): with severe insufficiency (SCAP: 13 patients, lipase output <10% NV and pancreo-lauryl test <20%) and with mild insufficiency or normal secretion (NCAP: 17 patients).

Results. The SCAP patients present a significantly greater gastric acidification in the 2 postprandial periods (m: 1.4, 1.6; range: 1–4, 1.2–3.8) compared to C (m: 2.9,

2.6; range: 1.2–4.7, 1.2–3.9; $p<0.001$) and NCAP (m: 2.3, 2.2; range: 1.6–4.5, 1.1–4.5, $p<0.05$).

Conclusion. Gastric acidity, exocrine pancreatic insufficiency and maldigestion appear to be closely correlated in CAP.

14 Comparison of Pancreatic Duct Morphology and Exocrine Function

T. Bozkurt, U. Braun, S. Leferink, G. Lux
Department of Internal Medicine and Gastroenterology, Community Hospital of Solingen, FRG

A comparative, prospective analysis between pancreatic morphology and exocrine function was performed in 49 patients (39 men, 10 women; age 23–79, mean 47 years). All patients underwent ultrasound (US), computed tomography (CT), ERP and secretin-caerulein test. The classification of pancreatogram findings was based on Cambridge terminology. In 10 patients no pancreatic duct changes were found by the ERP. Slight (Cambr. I), moderate (Cambr. II) and marked ductal changes (Cambr. III) were found in 10, 13, and 16 patients, respectively. In patients with Cambr. I, II, and III a correlation of pancreatic functional impairment and US/CT changes was found in 50, 69 and 100%. In patients with normal pancreatogram only 1 patient had pancreatic functional impairment. Patients with slight pancreatic duct changes presented 50% dissociated and 50% global pancreatic insufficiencies, while patients with moderate pathologic duct morphology showed 38% dissociated and 62% global functional impairments. All patients with severe pancreatic duct changes had global pancreatic insufficiency. In comparison with normal subjects, volume, enzyme and bicarbonate secretion was reduced by about 35, 55, and 80% in patients with Cambr. I, II, and III duct morphology, respectively. In all groups bicarbonate and lipase seemed to be the most sensitive parameters. Preliminary results of this study assume that ERP findings with ductal changes of Cambr. III make pancreatic function tests unnecessary.

15 Exocrine Pancreatic Function after Recovery from Necrotising Pancreatitis

T. Bozkurt[a], D. Maroske[a], G. Adler[b]
Departments of Gastroenterology and General Surgery, Universities of [a]Marburg and [b]Ulm, FRG

The aim of the study was to investigate the exocrine pancreatic function in patients who had recovered from necrotising pancreatitis. We investigated 65 patients who had recovered from their first attack of necrotising pancreatitis. Exocrine pancreatic function was observed after endogenous stimulation using the Lundh test meal. A group of 37 patients was submitted to a Lundh test 18 months after recovery from acute pancreatitis. Of these patients, 5% showed a normal pancreatic response, while slight-to-moderate insufficiency occurred in 87%. Severe pancreatic insufficiency was observed in 8% of the patients. In a second group of 28 patients, 3 follow-up studies were performed between 4 weeks and 12 months after recovery from acute pancreatitis. After 4 weeks, 25% of the patients had slight-to-moderate insufficiency, while 75% suffered from severe impairment of pancreatic function. Twelve months after recovery, only 15% of the patients showed a normal pancreatic response, while 85% still had pancreatic insufficiency of variable severity. Individual pancreatic enzymes varied considerably during the course of recovery in the follow-up period. The results of the present study show that recovery from necrotising pancreatitis is associated with persistent global or dissociated functional impairment of the exocrine pancreas in 85–90% of the patients.

16 Adjuvant Omeprazole vs. Cimetidine Treatment during Enzyme Replacement Therapy in Pancreatic Insufficiency

M.J. Bruno, E.A.J. Rauws, F.J. Hoek, G.N.J. Tytgat
Department of Gastroenterology, Academic Medical Centre, Amsterdam, The Netherlands

Aim of the Study. We compared the efficacy of omeprazole (OME; 60 mg, once daily before breakfast) with cimetidine (CIM; 400 mg, 3 times daily 30 min before main meals) as adjuvant therapy to reduce faecal fat excretion in patients with chronic pancreatic insufficiency treated with pancreatic enzyme supplements.

Patients and Methods. In a cross-over double-blind study 10 patients were analyzed (9 men, 1 woman). The median age was 64 years (range: 36–68). During weeks 1 and 4, patients received single enzyme replacement therapy (cotazyme forte; 6000 FIP units lipase/tablet). Patients used 6 tablets during each of the 3 main meals daily. During weeks 2–3 and 5–6, OME or CIM was supplemented in a cross-over fashion according to randomization. Seventy-two hours' faecal fat excretion was determined at the end of weeks 1, 3 and 6 (100 g/day fat diet). Faecal fat was determined by the method of Van de Kamer. Statistical analysis was performed using the Wilcoxon matched pairs signed rank sum test (two-tailed).

Results. During enzyme replacement therapy the mean faecal fat excretion was 33.1 g/day (SD: 14.0), during adjuvant OME 13.0 g/day (SD: 6.1) and during adjuvant CIM 20.9 g/day (SD 11.2). Both adjuvant OME and CIM treatments significantly decreased faecal fat excretion compared with single enzyme replacement therapy ($p<0.01$). Although OME showed a trend towards a more favourable decrease of faecal fat excretion compared with CIM, variation among measurements was too big, or the number of patients too small, to show a statistically significant difference ($p=0.10$).

Conclusion. Both adjuvant antisecretory OME and CIM treatment significantly reduce the amount of fat in faeces during pancreatic enzyme replacement therapy in patients with chronic pancreatic insufficiency. No statistical difference between adjuvant OME or CIM in the efficacy to reduce faecal fat excretion was found.

17 Morphofunctional Study of the Main Pancreatic Duct (MPD) by US-S Test Before and After Endoscopic Sphincterectomy (ES) in Patients with Acute Relapsing Pancreatitis (ARP)

G. Cavallini[a], *G. Angelini*[a], *L. Rigo*[a], *P. Bovo*[a], *B. Vaona*[a], *M.P. Brunori*[a], *V.DiFrancesco*[a], *L. Frulloni*[a], *D. Sgarbi*[a], *P. Pederzoli*[b], *G. Talamini*[a]
[a]Clinica Medica and [b]Clinica Chirurgica, University of Verona, Italy

More than 70% of patients with ARP present defective outflow of pancreatic secretion, as shown by pathological responses to the Ultrasonography-Secretin (US-S) test. The aim of this study was to assess whether endoscopic sphincterectomy (ES) of the major or minor papilla is capable of affecting the results of the test in ARP patients with pathological US-S test findings. Eight patients (4 men, 4 women; mean age: 47 ± 17 years) with recurrences of acute pancreatitis (3.2 ± 1.7 relapses over 4.3 ± 2.7 years; 3 pancreas divisum, 1 odditis, 4 idiopathic recurrent pancreatitis) underwent ultrasonographic measurement of the MPD every 5 min for 60 min after maximal stimulation with secretin (S) 1 IU/kg both before and 1 month after a single ES session.

Results. Before ES, the diameter of the MPD was abnormally dilated throughout the test with values significantly higher than those in controls (Mann-Whitney U test, $p<0.05$). After ES, MPD diameters showed a significant reduction compared to baseline values (Wilcoxon test, $p<0.05$), with a curve trend closer to that of the controls. The US-S test results showed improvement after ES in all patients. No recurrences of acute pancreatitis were observed in the post-ES, follow-up period (1–18 months). We conclude: (1) that pancreatic outlet obstruction in ARP may be corrected by ES, and (2) that the US-S test permits objective evaluation of the efficacy of ES.

18 Origin of Oxygen-Free Radicals in Acute Experimental Pancreatitis

D. Closa[a], *O. Bulbena*[b], *E. Gelpí*[b], *L. Fernández-Cruz*[a]
[a]Department of Surgery, Hospital Clinic, University of Barcelona, and [b]Molecular Pathology Unit, CID-CSIC, Barcelona, Spain

Oxygen-free radicals and prostaglandins are implicated in the physiopathology of acute pancreatitis, although their mechanisms of action remain unclear. We have studied the effect of administration of 16,16-dimethyl-PGE_2 on oxygen-free radical production in acute pancreatitis.

Material and Methods. Five experimental rat (300 g BW) groups have been studied. Group I: control; group II: sodium taurocholate (3.5%) induced; group III: like group II with previous administration of 16,16-dimethyl PGE_2 (0.5 μg/kg); group IV: like group II with previous administration of indomethacin (20 mg/kg); group V: like group II with previous administration of superoxide dismutase (13 mg/kg). Five min after induction pancreas samples were taken for xanthine oxydase/dehydrogenase assay.

Results. In sodium taurocholate treated rats, xanthine dehydrogenase is completely converted to xanthine oxidase within the first 5 min with subsequent oxygen-free radical production, while in 16,16-dimethyl PGE_2 treated rats this enzymatic transformation does not occur. In the superoxide dismutase treated group xanthine oxidase activation is partially prevented.

Conclusion. These data suggest that xanthine oxidase is the main source of oxygen-free radicals, which contribute to extend the cellular damage in sodium taurocholate induced acute pancreatitis.

19 Extracorporeal Shock Waves Lithotripsy (ESWL) of Pancreatic Stones (PS) in Patients with Chronic Calcifying Pancreatitis (CCP). Preliminary Results

G. Costamagna[a], *A. Gabbrielli*[a], *M. Mutignani*[a], *V. Perri*[a], *D. Di Muzio*[b], *P. Caradonna*[b], *F. Crucitti*[a]
Departments of [a]Surgery and [b]Medicine, Catholic University of Roma, Italy

Increased pressure in the main pancreatic duct (MPD), secondary to duct stricture or intraductal calculi is a relevant cause of pain in patients with CCP. Pancreatic ductal drainage, in patients with CCP, can be achieved endoscopically by stone removal after endoscopic pancreatic sphincterotomy (EPS) and/or by insertion of pancreatic stent. The embedding of pancreatic calculi in the MPD prevented extraction in same cases. Good results have been reported by Cremer disintegrating pancreatic stones by ESWL with successful subsequent endoscopic extraction of the fragments. We report preliminary results in our first 5 patients.

Patients. From December 1991 to April 1992, 5 patients (5 men; mean age 45.8, range 31–72 years) suffering from severe painful CCP with impacted obstructing PS and upstream dilatation of the MPD were treated by ESWL. The etiology was alcoholic in 3 cases and idiopathic in 2. Two patients underwent previous surgery for CCP.

Method. ESWL was performed using electromagnetic lithotriptor from Siemens (Erlangen, FRG), the Lithostar which delivered shock with a power ranging from 13 to 19 kV utilizing for focusing a bidimensional X-ray system. Patients were sedated with an intravenous injection of 5–15 mg of diazepam. The mean number of sessions for patients was 2.2 (1–3). A mean number of 3500 shock waves (2500–5000) per session and 7700 (4800–9900) per patient were administered. EPS was performed before and immediately after ESWL in 4 and 1 case, respectively. Complementary endoscopic treatment was performed for biliary stricture in 2 patients (2–10 F biliary stents), for MPD stricture in 2 patients (2–10 F pancreatic stents), and for a head pancreatic cyst (cystoduodenostomy) in 1 patient.

Results. The procedures were well-tolerated by the patients. All the calculi were fragmented and a complete clearance of MPD with relief of pain and decrease of MPD diameter was obtained in all the patients. No morbidity correlated to the procedure was observed in 4 patients. One patient suffered from a sepsis of pancreatic origin, probably related to the endoscopic procedure.

Conclusion. These preliminary results show that ESWL is a safe and effective treatment for impacted PS of the MDP. ESWL combined with endoscopy may increase the success rate in the non-surgical treatment of patients with CCP.

20 Endoscopic Treatment of Chronic Calcifying Pancreatitis (CCP)

G. Costamagna[a], *M. Mutignani*[a], *A. Gabbrielli*[a], *V. Perri*[a], *P. Pianozza*[b], *M. Boscaini*[c], *P. Caradonna*[b], *F. Crucitti*[a]
Departments of [a]Surgery and [b]Medicine, Catholic University, [c]Lithotripsy Unit Pio XI, Rome, Italy

Obstruction of the main pancreatic duct (MPD) with secondary upstream hypertension and dilatation is a relevant cause of pain in patients with CCP. In patients selected on a morphological basis (distal stenosis of MPD with upstream dilatation, obstructive stones, extrinsic compression by cysts), it is possible to remove the obstruction endoscopically in order to restore a physiologic outflow of pancreatic juice.

Patients and Methods. Between November 1987 and April 1992, 131 patients with CCP have been examined by ERCP; 51 symptomatic patients (38.9%) selected on a morphological basis have been treated endoscopically (43 men, 8 women; mean age 49.9 years, 19–79). The etiology of CCP was alcohol in 36 patients and idiopathic in 15. Eleven patients had undergone previous surgery for CCP. Forty EPS were performed. A pancreatic stent was inserted in 27 patients; in 14 cases 1 or more biliary stents were inserted for symptomatic biliary obstruction. In 3 patients a self-expanding metal stent (Wallstent) was positioned (2 biliary and 1 pancreatic). Ten patients underwent ESWL for obstructive impacted stones. Endoscopic cystoduodenostomy was performed in 4 patients. In 2 patients with communicating cysts a complete drainage by a naso-pancreatic catheter inserted transpapillarily into the cysts was obtained. In 1 patient transpapillary drainage failed because of the inability to insert the guide-wire into the cyst.

Results. There was no mortality related to the endoscopic technique. The morbidity rate was 7.8% (1 acute pancreatitis after EPS with abscesses formation which required surgery and had a favourable outcome; 1 acute pancreatitis after endoscopic biliary sphincterotomy treated conservatively; 1 acute cholangitis treated by endoscopic drainage; 1 acute cholecystitis treated conservatively). All but 2 patients experienced an immediate disappearance of pain corelated to the reduction in the diameter of the MPD (in these 2 cases it was not possible to obtain a complete drainage of the pancreatic duct in 1 case and of a cyst in another). Three patients died during the follow-up: 2 of ictus cerebri and 1 of acute pancreatitis after an alcoholic abuse. Two patients dropped out. Three patients were excluded from the follow-up because of an incomplete treatment. One patient continued the alcoholic abuse. Three patients were operated on: 2 pancreaticojejunostomy and 1 duodenocephalopancreatectomy. The mean follow-up of the remaining patients is 22 months (1–51). Twenty-four pain relapses were recorded in 13 patients (33.3%) due to pancreatic stent obstruction (67.7%), to alcoholic abuse (12.5%), to migration of pancreatic stones from secondary ducts (8.3%) and to incomplete drainage after ESP (4.2%). Two patients (8.3%) developed a pancreatic pseudocyst after a painful relapse with patent pancreatic stent. The mean clogging time is 265 days (83–615) for the pancreatic and 122 days (107–157) for the biliary stents. Replacement of the obstructed pancreatic stent was followed by immediate decompression of the MPD and disappearance of pain. There was no case of stent occlusion in patients treated with self-expanding stents.

Conclusion. The endoscopic treatment of CCP in selected patients showed to be safe and effective in the treatment of pain in the short and medium time follow-up.

21 Endotoxin Exposure in Human Acute Pancreatitis Is Exacerbated by Interventional Radiological Techniques

P.J. Curley[a], *G.R. Barclay*[b], *M.J. McMahon*[a]
[a]Academic Unit of Surgery, University of Leeds, and [b]Scottish Blood Transfusion Service, Edinburgh, UK

Introduction. Endotoxaemia occurs in up to 100% of severe attacks of acute pancreatitis, is difficult to measure using conventional Limulus Lysate assays (LAL) and may account for the sepsis syndrome often seen after surgical and radiological intervention for pancreatic necrosis, abscess or pseudocyst.

Hypothesis. Operative and radiological interventions are accompanied by systemic endotoxaemia.

Methods. Thirty-two patients with acute pancreatitis (21 severe, 11 mild) were studied. Endotoxin and circulating IgG and IgM anti-endotoxin-core antibody levels were measured and compared to 1000 healthy blood donors. Seven patients had percutaneous drainage procedures and 1 patient underwent necrosectomy.

Results. Circulating endotoxin was detected in 7/21 (33%) of severe attacks and 0/11 mild attacks ($p>0.05$). There was no significant difference in levels of IgG antibodies between groups, but there was a significant decrease in IgM levels between mild and severe attacks compared to controls ($p<0.05$). In addition 7/8 patients had a fall in IgG immediately after intervention compared to 9/25 patients not undergoing intervention studied on consecutive days (χ-square $p<0.05$).

Conclusion. Exposure to endotoxin is common in acute pancreatitis, is poorly identified by LAL assay and is exarcerbated by manipulation of the pancreas.

22 Diet-Induced Murine Acute Pancreatitis Is Associated with Reduced Interleukin-2 Production and Increased Mortality from Endotoxaemia

P.J. Curley, K.H. Collins, M.S. Nestor, J.A. Mannick, M.L. Rodrick
Department of Surgery, Brigham, and Women's Hospital, Boston, USA

Introduction. Reduced interleukin-2 production is a fundamental intracellular abnormality of human and murine mononuclear cells after burns and blunt trauma. Upregulation of IL-2 production by administration of exogenous recombinant IL-2 can improve mortality from sepsis in a mouse model of burn injury. Infusion of endotoxin to healthy volunteers produces similar abnormalities and endotoxaemia is common in acute pancreatitis. The hypothesis tested in this study was that reduced IL-2 production occurs in mice with acute pancreatitis and may render mice susceptible to endotoxin.

Methods. Female Swiss-Webster mice (20–25 g weight) were fed a choline deficient diet supplemented with 0.5% DL-ethionine. Mice were sacrificed on days 3, 7 and 10 after introduction of the diet and splenic mononuclear cells were harvested and stimulated in vitro with Concanavalin A for 48 h after which IL-2 production was measured using bioassay. Another group of mice was challenged with 0.25 mg/100 g endotoxin intraperitoneally on day 10 and 3-day survival compared to normal mice.

Results. Interleukin-2 production was reduced compared to controls at all time points (33% day 3, $p<0.05$; 25% day 7, $p>0.05$; 39% day 10, $p<0.005$). Mortality following endotoxin challenge was 0% in control animals and 90% in animals with acute pancreatitis ($p<0.0001$).

Conclusion. Acute pancreatitis is associated with reduced mononuclear cell IL-2 production and markedly increased sensitivity to endotoxin. This model may be suitable for the investigation of immunomodulatory therapy in acute pancreatitis.

23 Circulating Mononuclear Cell Surface Antigen Expression in Human Acute Pancreatitis

P.J. Curley[a], *F. Lancaster*[b], *A.W. Boylston*[b], *M.J. McMahon*[a]
Academic Units of [a]Surgery and [b]Pathology, University of Leeds, UK

Decreased CD4+ lymphocytes (T-helper cells), decreased CD8+ cells (T suppressor cells) and reduced CD4/CD8 ratios have all been reported after endotoxin infusion, and after burn and blunt trauma in humans. Since endotoxaemia is an event common to human acute pancreatitis, burns and trauma, we hypothesised that abnormal cell surface antigen expression occurs in patients with pancreatitis.

Thirteen patients with severe acute pancreatitis and 10 with mild attacks were studied prospectively within 24 h of admission. Four of the severe group died and 6 of the survivors were restudied 3 months after the acute attack (severe/convalescent). Peripheral circulating mononuclear cells were isolated and flow cytometric analysis of CD4, CD8 and CD3 antigen expression performed.

Significant reductions in circulating CD3 ($p<0.001$), CD4 ($p<0.001$), CD8 ($p<0.001$) positive cells were seen in patients with severe acute pancreatitis compared to mild attacks and to controls. All parameters measured, with the exception of CD4/CD8 ratio, showed significant improvement in survivors of severe attacks ($p<0.005$; $p<0.01$; $p<0.05$).

	Severe	Mild	Control	Severe/ convalescent
CD3	29.8%	55.2%	62.2%	58.4%
CD4	11.4%	34.8%	38%	30.9%
CD8	9.1%	19.4%	23.4%	19.1%
CD4/CD8	1.24	1.9	1.5	1.32

Similar immunological abnormalities are present in patients with severe acute pancreatitis as are found in other forms of surgical disease. Patients with severe acute pancreatitis may be suitable for immunomodulatory treatment regimens aimed at improving T-lymphocyte function.

24 Effect of Platelet-Activating Factor (PAF) Antagonist on Oxidant-Antioxidant Balance in Acute Haemorrhagic Pancreatitis in the Rat

A. Dabrowski, A. Gabryelewicz, L. Chyczewski
Departments of Gastroenterology and Pathomorphology, Medical School, Bialystok, Poland

We have recently found serious, multiorgan disturbances of oxidant-antioxidant metabolism in early hours of Na-taurocholate induced acute pancreatitis (Na-TC-induced AP) in the rat. PAF is a potent inflammatory agent responsible for pancreatic oxidative stress which occurs early in the course of cerulein-induced acute pancreatitis. The aim of the present study was to investigate the influence of PAF-receptor antagonist (BN 52021)* on the development of Na-TC-induced AP.

Methods. Acute pancreatitis was induced in Wistar rats using a retrograde intraductal injection of 5% Na-taurocholate. It was immediately followed by i.v. injection of BN 52021 (10 mg/kg). Rats were sacrificed at 3 h of Na-TC-induced AP. Malondi aldehyde (MDA) concentration, as an index of oxidative injury, and sulfhydryl compounds (SH)-major nonenzymatic antioxidant, were measured in pancreatic, liver and lung tissue. Pathomorphology of the pancreas was studied by light microscopy.

Results. Na-TC-induced AP caused significant increase of MDA in pancreatic and lung tissue, accompanied by depletion of SH, by 40 and 33%, respectively. Only slight decrease of SH was observed in liver tissue. In rats treated with BN 52021, lower (by 56%) serum amylase and pancreatic MDA (by 53%) increase was found; there was no lung tissue SH depletion. BN 52021, however, did not protect against pancreatic SH diminution and oedema development; lung tissue MDA elevation was the same as in untreated rats. BN 52021 diminished the features of blood vessel necrosis as well as leukocytes margination and infiltration within pancreatic tissue.

Conclusion. BN 52021 exerts only partial protecting effect against pancreatic and lung oxidative injury in Na-TC-induced AP in the rat. It is likely that this effect depends on the inhibition of PAF-mediated oxidant generation by phagocytes.

*Gift of Dr. P. Braquet.

25 TPS: An Useful Index in Diagnosing and Monitoring Pancreatic Cancer?

G. Del Favero[a], D. Basso[b], C. Ferrara[a], T. Meggiato[a], P. Fogar[a], M.P. Panozzo[a], M. Plebani[b], R. Naccarato[a]
[a]Cattedra di Gastroenterologia e [b]Istituto di Medicina di Laboratorio, Università degli Studi di Padova, Italy

TPS has been recently proposed as an useful index of tumor growth. The aim of this study was to evaluate TPS clinical utility in assessing patients with pancreatic cancer (PC) both at diagnosis and follup-up. Results were compared with those obtained with TPA and CA 19-9. We studied 40 patients with PC. Staging was as follow: stage 1: 5 patients; stage 2: 9 patients; stage 3: 14 patients; stage 4: 12 patients. Tumor volume was evaluated by means of tomography. Pathological values of CA 19-9, TPA and TPS were found in 37/40, 23/40 and 31/40 patients. Higher mean values of TPA ($F=2.98$, $p<0.05$) and of CA 19-9 ($F=5.33$, $p<0.005$) were found in those patients with an advanced PC (stage 4) as compared to patients with a local spread of the tumor (stages 1, 2 and 3). No significant difference among stages was detected for TPS. In particular high levels were found in all the 5 stage – 1 patients. No correlation was found between the values of all the 3 markers and tumor volume. None of the 3 indices was able to predict survival time. However, differently from CA 19-9, TPA and TPS were in the normal range in the 6 patients who survived more than 10 months. Tumor volume did not predict survival time, which was shorter in PC patients with liver metastases ($Z=1.98$, $p<0.05$). TPA and TPS significantly decreased after surgical treatment (both radical or palliative) ($t=1.88$, $p<0.005$, respectively), while no significant trend was observed in non-treated patients.

Conclusions. (1) TPS seems to be a sensitive index of PC especially in detecting early tumors; (2) high levels of this antigen as well as of TPA, but not of CA 19-9, seem to be related with a shorter survival time; (3) however, an advanced tumor is the main factor which can influence survival; (4) surgical treatment is associated to reduced levels of TPS and TPA, probably because of jaundice relief, able to influence serum glycoproteins, like TPA and probably TPS, catabolism.

26 **C Peptide Pattern in Patients with Pancreatic Cancer**

G. Del Favero[a], *P. Fogar*[a], *T. Meggiato*[a], *D. Basso*[b], *M. Plebani*[b], *G. Briani*[a], *M.P. Panozzo*[a], *C. Ferrara*[a], *R. Naccarato*[a]
[a]Cattedra di Gastroenterologia, [b]Istituto di Medicina di Laboratorio, Università degli Studi di Padova, Italy

In the present study we investigated: (1) the pattern of some hormones involved in glucose homeostasis in patients with pancreatic cancer (PC) with (HPC) or without hyperglycemia as compared to patients with hyperglycemia of well-known origin and (2) the role of different aspects of PC in influencing the variations of the studied hormones.

In the sera of 42 patients with PC (32 having and 10 not having hyperglycemia), 8 with type I diabetes mellitus (DM) and 22 with diabetes secondary to liver cirrhosis (LC: 9 cases) or chronic pancreatitis (CP: 13 cases), C-peptide, glucagon and growth hormone (GH) were determined. C-peptide significantly differed among groups ($F=3.76$, $p<0.01$). Low levels (<0.5 μg/l) were found in CP and DM in 68 and 100% of the cases, respectively. GH did not significantly vary among groups. In HPC glucagon levels were in the normal range, while higher mean values were recorded in LC ($p<0.05$) as compared to PC and CP patients. No correlation was found among the 3 hormones. In HPC C-peptide and glucagon were found to vary in a consensual manner ($p<0.0001$). Neither tumor volume (<4> cm) or the presence (16 patients) or absence of liver metastases, or a family history of DM, or a pre-existing DM (diagnosed more than 48 months before the onset of PC) (9 patients) influenced C-peptide; higher levels of GH ($p<0.05$) and glucagon ($p<0.005$) were found in HPC patients with than those without liver metastases. We may conclude: (1) HPC patients presented 2 distinct patterns of C-peptide serum levels, one normal/low and another high; (2) the variations of C-peptide in these patients cannot be explained on the basis of tumor volume and spread; (3) GH and glucagon do not seem to be involved in determining hyperglycemia in PC patients; (4) a peripheral insulin resistance in some patients with PC may be hypothesized.

27 **Bile Infection in Acute Gallstone Pancreatitis**

B. De Waele[a], *S. Lauwers*[b], *G. Willems*[a]
Departments of [a]Surgery and [b]Microbiology, VUB University Hospital, Brussels, Belgium

Bacterial infection of bile is thought to be a key factor in the pathogenesis and clinical course of acute biliary pancreatitis (ABP). The aim of the present study was to determine the microbiologic profile of biliary infection in ABP and the sensitivity of the infective organisms to currently used antibiotics. Forty-eight patients with proven ABP entered the study. At operation, bile from the gallbladder and/or common duct was obtained for cultures.

Results. In 20 of the 48 patients (41.7%) microorganisms were cultured from bile while in 28 patients no growth was obtained. In 8 patients more than one organism was isolated. Sixteen Enterobacteriaceae – 10 Escherichia coli, 4 Klebsiella sp., 1 Proteus mirabilis and 1 Citrobacter freundii – were recovered in 11 patients. Gram-positive cocci were found in 9 patients: Enterococcus sp. in 6, streptococci in 2 and a coagulase-negative staphylococcus in 1 patient. Yeasts were isolated as the sole organism in 2 patients and Clostridium perfringens was found in 2 patients. The isolated bacteria were quite sensitive to antibiotics: all Enterobacteriaceae were sensitive to netilmicin, 9 of 16 were sensitive to ampicillin (7 of 10 E. coli were ampicillin S) and 15 of 16 were sensitive to cefazolin and trimethoprim sulphamethoxazole.

Conclusion. In our patient population acute gallstone pancreatitis is frequently associated with biliary infection by microorganisms which are quite sensitive to currently used antibiotics.

28 **β-Cell Function in Chronic Pancreatitis. Enzyme Replacement Effect**

D. Di Muzio[a], *P. Pianozza*[a], *P. Caradonna*[a], *M. Mutignani*[b], *G. Costamagna*[b]
Departments of [a]Internal Medicine, Chair of Metabolic Diseases, and [b]Surgery, Catholic University, Rome, Italy

Exocrine pancreatic deficiency has a significant effect on β-cell function. The maximum insulin secretory capacity is found to be reduced. Postprandial GIP and CCK responses are impaired in chronic pancreatitis. The oral pancreatic enzymes addition can improve GIP and CCK response to meal. The positive effect of these hormones on insulin secretion are known. Our study purpose was to answer the question whether or not a relationship exists between oral administration of pancreatic enzymes and β-cell function. Six male and 2 female subjects suffering from severe chronic pancreatitis (ECRP diagnosis) were studied. Glucose tolerance and fecal fats were normal. The β-cell secretion was assessed by measuring C-peptide after intravenous glucagon stimulation. The test was performed before and one month after enzyme addition (enteric-coated microsphere: lipase 130,000 units/day). The patients showed a blunted C-peptide output. After pancreatin medication C-peptide response improved but not significantly. Glycemic response was unaltered. Our results suggest that the improvement of nutrient absorption with related 'incretin' effect resulted in a favourable insulinotropic activity.

Glucose and C-peptide values after I.V. glucagon

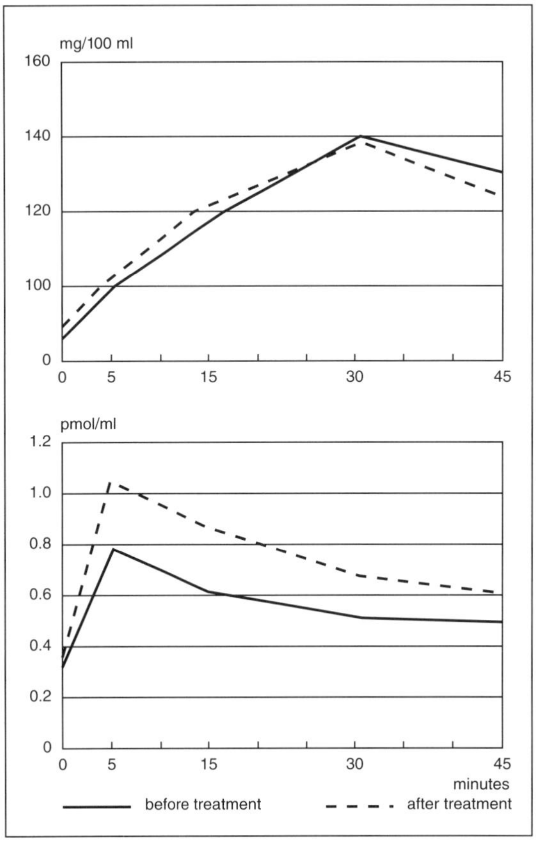

29 Growth Factors, Cytokines and Adhesion Molecules in Pancreas Carcinoma Cells

W. Dippold, H. Kaulen, W. Schwäble, M. Kerlin, K.-H. Meyer zum Büschenfelde
I. Med. Klinik, Johann Gutenberg University, Mainz, FRG

Recently, growth factors, cytokines and adhesion molecules have been characterized by complementary DNA cloning. We could demonstrate that epidermal growth factor (EGF), interleukin-3 and granulocyte-macrophage colony-stimulating factor stimulate growth of pancreas carcinoma cells. The growth stimulation was independent of EGF-receptor mRNA abundance. Subsequently, we have examined the expression of interleukin-6, a major regulator of acute-phase protein synthesis, also known as hepatocyte stimulating factor and which acts as a growth factor in plasmocytoma cells. In addition, the occurrence of the cell adhesion molecule ICAM-1 was studied on mRNA and protein level in 6 surgically removed pancreatic tumors and 8 pancreatic tumor cell lines.

Immunohistochemistry revealed a varying percentage of ICAM-1 positive pancreas tumor cells while normal pancreatic tissue was not stained. Three pancreatic tumor cells expressed ICAM-1 constitutively. In 4 ICAM-1 negative pancreas cancer cell lines ICAM-1 expression could be induced remarkably on mRNA and protein level by incubating the cells in the presence of inflammatory cytokines (TNF-α, IFN-γ, IL-1β). Regulation of the ICAM-1 expression in those cells occurred at the transcriptional and post-transcriptional level. 2/8 pancreas carcinoma cell lines testes produced IL-6 mRNA and protein and expressed IL-6 receptor mRNA as shown by PCR.

Presently, we investigate whether these findings confer to the aggressiveness of pancreas cancer.

30 Interdigestive Exocrine Pancreatic Function and Motilin Release in the Early Phase of Acute Pancreatitis (AP) in Humans

J.E. Domínguez-Muñoz[a], *O. Pieramico*[a], *M. Büchler*[b], *P. Malfertheiner*[a]
Departments of [a]Internal Medicine and [b]Surgery, University of Ulm, FRG

The purpose of the present study was to analyze interdigestive exocrine pancreatic secretion and the release of pancreatic polypeptide (PP) in the early phase of AP in humans and their relationship to the release of motilin.

Material and Methods. Six patients with AP (4 biliary, 1 alcoholic and 1 idiopathic AP) were studied within 72 h after the onset of symptoms. Diagnosis of AP was based on clinical criteria, laboratory parameters (serum pancreatic enzymes >3 times upper limit of normal), and findings on CT. On angio-CT, 2 patients had necrosis of $\leq$30% of the pancreas and 4 patients had edema alone. In the 24 h preceding the study, solutions containing exclusively electrolytes were administered intravenously. Twelve healthy controls matched by both sex and age were studied. Interdigestive pancreatic secretion was studied by standard duodenal intubation technique (PEG 4000 15 g/l, 3 ml/min). Output of trypsin, chymotrypsin, and amylase were calculated every 15-min period for 4 h. Plasma PP and motilin were measured (RIA) at 15-min intervals. Data are expressed as integrated enzyme output and hormone release per h.

Results. The procedure was well tolerated by all patients. Results in patients with AP (individual data) and in controls (median, 95th percentile) are shown in the figure. Only 1 patient showed impaired pancreatic secretion during a second attack of necrotizing AP, pancreatic secretion and release of PP and motilin maintained the coordinated and cyclical pattern.

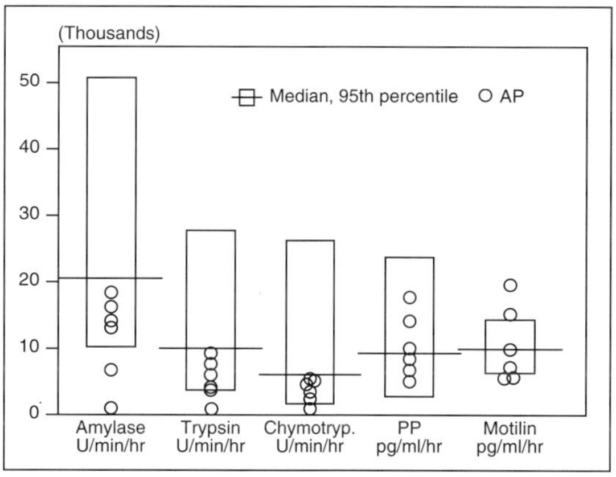

Conclusions. Interdigestive exocrine pancreatic function (pancreatic secretion and PP release) in the early phase of mild to moderate AP in humans appears to be normal. This is in contrast to what could be expected from data on experimental AP in animals.

31 Oxygen Radicals in Acute Haemorrhagic-Necrotizing Pancreatitis

S. Eisele, M.H. Schoenberg, M. Büchler, M. Younes[a], *H.G. Beger*
Department of General Surgery, University of Ulm; [a]Max-von-Pettenkofer-Institut, BGA, Berlin, FRG

Oxygen radicals (OR) play an important role in mediating tissue damages in the acute edematous pancreatitis. There are however conflicting results in regard to the acute haemorrhagic-necrotizing pancreatitis (NP). Aim of our experimental study was to reevaluate the participation of OR in acute NP.

Methods. In rats NP was induced by retrograde injection of 5% sodium-taurocholate into the pancreatic duct. The injection pressure did not exceed 15 cm H_2O. After 0.5, 1, 3.5 and 12 h the pancreas organ was removed for light microscopy. Blood samples were drawn for serum enzyme measurement. Forty rats were treated before induction of NP with the radical scavengers (FRS) superoxide dismutase (SOD, dosage 150,000 U/kg/h) and catalase (CAT), dosage 400,000 U/kg).

Results. In the untreated group first morphological changes became apparent after 0.5 h (edema). Infiltration of neutrophils was first seen after 3.5 h. After 3.5 and 12 h, 51±9% and 88±14% of the acinus cells were necrotic. Serum amylase and lipase increased from 54±32 up to 50.610±8.936 U/l and from 72 ±27 to 19.630±5.559 U/l respectively after 12 h. After 12 h, 8 out of 14 rats died. In the animals treated with SOD and CAT, serum enzyme levels increased significantly less (amylase: 0.5 h, $p<0.0021$; 12 h, $p<0.0055$; lipase: 0.5 h, $p<0.012$; 12 h, $p<0.0122$). Concomitantly the tissue damages were significantly lowered (necrotic acinus cells: 14±7% after 3.5 h and 44±7% after 12 h, $p<0.0001$). 2 out of 10 rats died after 12 h.

Conclusions. OR seem also to play a pivotal role in the early phase of NP. Treatment with FRS significantly diminished the increase of serum enzyme levels, the rate of cell damage and mortality.

32 Secretory Immunoglobulin A Levels in Pure Pancreatic Juice

J. Emmrich, P. Conradi, M. Seyfarth, S. Liebe
Department of Medicine and Institute for Clinical Immunology, University of Rostock, FRG

Secretory immunoglobulin A (sIgA) is the most predominant class of immunoglobulins present in certain external secretions (saliva). The purpose of this study was to compare sIgA levels in pure pancreatic juice (PPJ) from patients with and without chronic pancreatitis (cP). The measurement especially of sIgA is affected by the presence of proteolytic enzyme activity in PPJ.

Methods. PPJ was obtained from 17 patients (9 patients with and 8 patients without cP) by catheterization of the main pancreatic duct during endoscopy and after stimulation of secretion by sequential injection of secretin (1 unit/kg) and pancreozymin (1 unit/kg). We could show that PPJ is activated mostly to a limited degree. Therefore it was necessary to search for potent inhibitors. We evaluated benzamidine, aprotinin, phenyl-methyl-sulphonyl-fluoride (PMSF), tosyl-lysin chloromethyl ketone (TLCK), diphenyl-carbamyl chloride (DPCC) and combinations of these. sIgA was estimated by ELISA method. A monoclonal antibody against human secretory component was coated in microculture plates. A peroxidase labelled anti-human-IgA serum was then used.

Results. To inhibit proteolytic enzyme activity we used a combination of TLCK and DPCC. The highest value of sIgA was 20 ng/ml in 9 patients without cP. Levels of sIgA were significantly elevated in patients with cP (435±210 ng/ml).

Conclusions. A precondition for exact analysis of PPJ is the inhibition of proteases. The most potent combination of inhibitors was TLCK and DPCC. sIgA levels were elevated in PPJ of patients with cP. This may reflect a local immunological reaction in the pancreas during cP.

33 Tumor Necrosis Factor Production in Patients with Multiple Abscesses Following Necrotizing Pancreatitis

G. Farkas[a], *Y. Mándi*[b], *S. Hégely*[b], *I. Béládi*[b]
[a]Department of Surgery, and [b]Institute of Micobiology, A. Szent-Györgyi Medical University, Szeged, Hungary

Sepsis and septic shock are the most frequent complications of pancreatic abscess. Tumor necrosis factor (TNF) plays a pivotal role in the initiation of septic syndrome. The aim of our study was therefore to investigate which of the TNF-related laboratory tests correlates with the severity of the illness or the lethal outcome of this type of sepsis. The serum TNF level was determined in 20 patients with multiple abscesses following necrotizing pancreatitis in a bioassay using the TNF sensitive WEHI 164 cell line and in TNF EIASA (Medgenix®) tests. The in vitro TNF-producing capacity of leukocytes was checked by LPS or by *Staphylococcus* induction. In all patients the surgical treatment consisted of wide-ranging necrosectomy combined with continuous widespread washing and suction drainage. Detectable TNF levels were in the sera of 30% of the patients. No correlation was observed between the serum TNF concentration and the severity of the illness. The in vitro TNF-producing capacity was higher in the patients in the study group, but it decreased on subsequent days, especially in fatal cases (3 patients). The transient loss of TNF production proved to be reversibly correlated with recovery. Our results suggest that determination of the TNF-producing capacity of the leukocytes might be more informative than measurement of the serum TNF level in evalution of the severity and prognosis of sepsis.

34 Bombesin Inhibits Growth of a Ductal Pancreatic Adenocarcinoma in Nude Mice But Not in Hamsters

A. Farré, P. Benítez, N. Nabau, C. Carcasona, C. Vidal, G. Capellá, F. Lluís, C.M. Townsend Jr.
Lab. Invest. Gastrointestinal, Hospital de Sant Pau, Barcelona, Spain; Department of Surgery, Galveston, TX, USA

Bombesin (BBS) is a peptide that stimulates growth of both mouse Swiss 3T3 fibroblasts and human small cell lung cancer. The H2T cells, a hamster ductal pancreatic adenocarcinoma cell line, possess specific binding sites for BBS.

Aims. To investigate the effect of chronic BBS treatment on growth of H2T cells inoculated in two experimental animal models.

Methods. In experiment 1, H2T cells were inoculated in cheek pouches of hamsters. Control group (n=8) received 0.1% BSA in saline, treated group (n=10) BBS 5 μg/kg. Hamsters were sacrificed at day 35. In experiment 2, athymic nude mice were inoculated with H2T cells subcutaneously. Control group (n=10) received 0.1% BSA in saline, treated group (n=10) BBS 5 μg/kg. Mice were sacrificed at day 32.

Results. Experiment 1, in hamsters no differences were observed in tumor and in pancreatic weight. Experiment 2, results in mice are depicted in the table (mean ±SEM; * = $p<0.05$; % = mg/100 g BW).

Treatment with BBS, at the dose tested, did not modify tumor and pancreatic growth in hamsters. By contrast, in nude mice treatment with BBS inhibited tumor growth, whereas it stimulated pancreatic growth.

Conclusions. (1) The finding of both trophic (in pancreas) and inhibitory (in tumor) responses suggests that the effect of BBS, or of peptides released by BBS (i.e. gastrin, CCK), are receptor-specific with a different effect in nor-

(Table to Abstract 34)

	Tumor		Pancreas	
	control	BBS	control (%)	BBS (%)
Area (mm²)	85.8 ±9.5	37.6 ±6.1*	–	–
Weight (mg)	244 ±43	106 ±25*	0.92±0.06	1.15±0.5*
DNA (mg)	0.69±0.09	0.33±0.06*	1.35±0.06	1.72±0.1*
RNA (mg)	0.86±0.2	0.51±0.07	15.5 ±0.75	19.1 ±1.14*
Protein (mg)	19.6 ±4.27	7.82±2.33*	145 ±8.87	194 ±15.3*

Mean±SEM; *$p<0.05$; % = mg/100 g BW.

mal and in neoplastic tissue. (2) Direct and indirect mechanisms and the dose of BBS administered may be involved in growth of normal and neoplastic pancreatic tissue.

35 Non-Immunologic Graft Pancreatitis in Pancreas Transplantation

L. Fernández-Cruz, R. Gilabert, L. Sabater, M.J. Ricart, A. Sáenz, E. Astudillo
Department of Surgery, Hospital Clinic, University of Barcelona, Spain

Non-immunologic graft pancreatitis in pancreas transplantation may lead to the loss of a well functioning graft. Their differentiation from graft rejection is important, since different treatments should be given. The aim of the study was to determine the etiology, impact on endocrine/exocrine function, complications and treatment of non-immunologic graft pancreatitis in pancreas transplantation.

Patients and Methods. Records of 52 whole-organ pancreas graft with urinary drainage were reviewed, and 23 episodes of non-immunologic graft pancreatitis were detected in 10 recipients. Diagnosis was based on clinical criteria, laboratory data, duplex-Doppler ultrasound and response to treatment.

Results. Twenty three episodes of non-immunologic graft pancreatitis were documented at a mean time post-transplant of 274±53 days. Pain in the graft site was found in 17 episodes, body temperature slightly elevated in 10, and dysuria in 7. Laboratory data showed an increase in serum amylase and lipase levels (1230±209 UI/1291±451 UI), without reduction in their urinary excretion. Ultrasound was pathologic in 10 episodes disclosing peripancreatic fluid in 10, parenchyma inhomogeneity in 3, and Wirsung duct dilatation in 1. No changes were observed in pulsed-Doppler study. Bladder dysfunction and/or urinary outflow obstruction was detected in 16 episodes, with simultaneous urinary tract infection in 10. Cytomegalovirus (CMV) infection was documented in 4 episodes. Bowel rest and bladder catheterization was effective to relieve the symptoms and to normalize the serum pancreatic enzymes in most of the episodes. Cystography showed a duodenal stump/bladder anastomotic leak in 3 patients, one of them with a duodenocolic fistula. Exocrine graft function became normal in 21 episodes. Two patients died (multiple organ failure in severe pancreatitis). Endocrine graft function was not impaired in 22 episodes.

Conclusion. The etiology of non-immunologic graft pancreatitis is in most cases unknown. Urinary catheterization was an effective treatment in 91% of episodes. Our results indicate that bladder dysfunction, causing reflux into the pancreas or temporary duct obstruction, and other factors such as urinary tract infection or CMV infection could be the cause of non-immunologic graft pancreatitis.

36 Ribonuclease Expression in Pancreatic Adenocarcinoma Cell Lines

E. Fernández, M. Ribó, R. Llorens, C.M. Cuchillo
Institut de Biologia Fondamental, Universitat Autònoma de Barcelona, Spain

The characterization of secreted ribonuclease from human pancreas and pancreatic adenocarcinoma cell line Capan-1 has shown that the two enzymes have identical amino acid sequence and kinetics parameters, and similar glycosylation.

Ribonuclease from human pancreas has hybrid and complex sugar chains N-linked to the three Asn target sites. The carbohydrate analysis is similar for tumoral and normal enzyme and showed high fucose contents indicating a possible Lewis x/y or SSEA-1 blood group determinant.

The presence of the messenger RNA for pancreatic ribonuclease in this cell line was shown by Northern blot. Other typical acinar enzymes were also monitored, as amylase and elastase, but no signal was detected. In contrast, typical ductal enzymes as carbonic anhidrase II and cytokeratin were expressed. Other authors reported trypsinogen expression in this cell line [1], another typical acinar enzyme. The analysis of ribonuclease expression in different human tissues by Northern blot is in progress.

Pancreatic ribonuclease is also secreted from the cell line Panc-1. Though, a different pattern of isoenzyme secretion was observed.

The evidence presented supports Reddy's work [2] about the involvement of acinar cells in pancreatic adenocarcinoma and renews the interest in acinar markers as ribonuclease.

1 Miszczuk et al., FEBS 1991;294:3,175–178.
2 Scarpelli et al., Cancer Cells 1991;3,7:275–279.

37 Neuropeptide Gene Expression and Neuroimmune Links in Chronic Pancreatitis

T. Fink[a], M. Büchler[b], P. Di Sebastiano[b], E. Weihe[a], M. Schäfer[a], H. Friess[b], H.G. Beger[b]
[a]Department of Anatomy, University of Mainz;
[b]Department of Surgery, University of Ulm, FRG

The reason for the generation and perpetuation of inflammation and pain in chronic pancreatitis is unclear. As pancreatic nerves from chronic pancreatitis are increased in their number and diameter and show damage to the perineurial sheath a neuronal pathogenetic factor in pancratitis has been suggested. Subpopulations of presumed sensory nerves containing the neuropeptides substance P and calcitonin gene-related peptide show signs of neuroproliferation and form close spatial relationships with inflammatory cells. In the view of the pronociceptive and immunomodulatory actions of these neuropeptides we postulated the possible significance of a neuroimmune link as a sustaining factor of pain and inflammation in this disease. In continuation of these observations, immunohistochemistry and in situ hybridization were used (1) to characterize the immune cells that are targets of peptidergic nerves, (2) to determine neuropeptide gene expression in intrinsic pancreatic neurons and their changes in pancreatitis, (3) to evaluate neuropeptide gene expression in inflammatory cells. Specific populations of nerves co-containing varied patterns of neuropeptides (substance P, VIP, PHI, NPY and opioid peptides) form close spatial relationships with specific subsets of immune cells that are strikingly increased in chronic pancreatitis. Neuro-T-cell contacts predominate. Intrinsic neurons express pro-enkephalin and tachykinin mRNAs that appear to be upregulated in pancreatitis. There was preliminary evidence of increased expression of neuropeptide mRNA in inflammatory cells occupying large areas of the inflamed pancreas. Changes in peptide gene expression in intrinsic and extrinsic pancreatic innervation and in inflammatory cells may reflect that neuroimmune mechanisms are involved in the pathogenesis and perpetuation of chronic pancreatitis.

(Supported by the BMFT.)

38 Routes of Pancreatic Enzyme Secretion in the Anaesthetized Rat

L.J. Fletcher, H. Seehra, H. Hadley, R.M. Case
Department of Physiological Sciences, University of Manchester, UK

Circulatory levels of pancreatic enzymes may originate directly via an endocrine route or indirectly via the lymphatic system. We have investigated this problem by studying the influence of both optimal and hyperstimulation on the secretion of newly synthesized and stored pancreatic enzymes.

Pancreatic juice, serum (from the portal vein) and lymph (from the superior intestinal lymphatic trunk) were collected from male Spague-Dawley rats, anaesthetized with an alphadolone/alphaxalone mixture. Animals were injected with a mixture of tritiated amino acids either 48 h before or immediately prior to the experiment. Samples were analyzed for amylase and TCA-insoluble radioactivity.

Optimal stimulation (0.5 μg/kg/30 min) with cholecystokinin-octapeptide (CCK-8) caused a significant rise in pancreatic juice and lymph amylase concentrations. Following hyperstimulation (5 μg/kg/30 min) both lymph and serum amylase concentrations increased significantly but a significant reduction in pancreatic juice amylase concentration was observed.

Optimal stimulation caused a gradual increase in the release of newly synthesized protein into pancreatic juice but not in lymph or serum. Hyperstimulation resulted in a smaller output of newly synthesized protein in pancreatic juice accompanied by a small rise in serum.

These data suggest a constant release of enzymes into the lymphatic and circulatory systems in the absence of stimulation and during optimal stimulation. A direct endocrine route to the circulation appears to be operative following hyperstimulation which consists of newly synthesized protein.

39 Role of Lipase in the Regulation of Upper Gastrointestinal Functions in Man: Nutrient Specific Effects – Studies with THL, a New Highly Specific Lipase Inhibitor

M. Fried, W. Schwizer, K. Asal, C. Güzelhan, J. Jansen, C. Lamers, A. Blum, J. Gonvers, D. Hartmann
University Hospitals, Lausanne, and Hoffmann-La Roche, Basel, Switzerland, and Nijmegen, The Netherlands

The role of lipase in the regulation of upper gastrointestinal functions and its nutrient specificity are poorly understood. We therefore studied the effect of tetrahydrolipstatin (THL) – a new highly specific and potent lipase blocker – on gastric emptying (GE) and pancreatobiliary secretion after a triglyceride and a glucose/albumin meal.

Results. (Table to Abstract 39)

Meal	GE	LO	TO	BO	CCK
Intra	8,793±333	739±205	17.9±1.8	388±83	425±87
Intra/THL	6,984±550*	234±74*	13.5±1.9*	270±68	262±89*
Glu/Alb	4,859±507	275±23	19.0±2.7	111±29	359±49
Glu/Alb/THL	4,757±570	161±27*	18.2±1.9	144±48	292±63

* $p<0.05$.

Methods. 6 healthy volunteers were examined using the double indicator technique with a triple lumen tube in the duodenum and PEG4000 as a duodenal perfusion marker and 99mTc-DTPA (Tc) as a meal marker. 500 ml liquid intragastric meals (500 kcal) of (a) 10 g% Intralipid (triglycerides) or (b) 12.5 g% glucose/10 g% albumin were given with or without 200 mg THL in random order. Measurement ($x \pm SEM$) of GE (AUC; %gastric Tc), outputs of lipase activity (LO; KU·240 min), trypsin (TO; mEq·240 min) and bilirubin (BO; mmol·240 min. Plasma CCK (pM·240 min) was measured by a specific and sensitive RIA.

Conclusions. (1) Lipase has an important nutrient specific role in the regulation of gastric emptying and pancreatobiliary secretion in man by promoting the release of CCK. (2) In the presence of fat only, the diminished release of CCK by THL results in more rapid GE rates and lower outputs of pancreatobiliary secretions.

(SNF grant 32.25727.88.)

40 Immunohistochemical Localization of Superoxide Dismutase in Human Pancreas in Relation to Age

Q. Fu[a], *F. Marotta*[b], *D.H. Chui*[c], *M. Kobayashi*[a]
[a]Pathology Department, Tokyo Women's Medical College, Tokyo, Japan; [b]Liver Unit Crespi, Niguarda Hospital, Milan, Italy; [c]Neurology Department, N. Bethune University, Changchun, China

The aim of the present study was to try to define a map of pancreatic SOD in relation to age. Pancreatic samples were obtained from 28 nondiabetic cases who had been autopsied within 3 h after death: four 22–36-week-old embryos, three 3–28-day-old newborns, eight 5-months- to 6-year-old children and thirteen of 31–81 years. Samples were treated with polyclonal anti-human CuZn SOD and Mn SOD. TEM immunohistochemistry was done by protein A-colloid gold method and double staining with insulin, glucagon and somatostatin. In adults, a diffuse and strong immunoreactivity for CuZn SOD was shown in centroacinar cc., perineural spaces, smooth cc. and endothelial cc. of the vessels, but Mn SOD had a weak immunoreactivity with a scattered pattern present only in centroacinar and periductular cc. Strong immunoreactive particles for CuZn SOD were found in the chromosomes of β cc. (weakly in the cytoplasmic matrix and organelles) and in chromosomes and secretory granules in α and d cc. Rare Mn SOD+ve particles were detected in cytoplasmic matrix of α, β and d cc. Islet cc. hormones staining had a strong positivity starting from birth. However, CuZn SOD had a weak positivity in embryos and in newborns. Both SOD stainings showed an age-related increase in immunoreactivity which appeared only until 6 years old.

41 Separation of Human Pancreatic Juice and Purification of the Human Pancreatitis-Associated Protein by Fast Protein Liquid Chromatography (FPLC)

M.J. Fuchs[a], *J.L. Iovanna*[b], *M. Büsing*[c], *U. Hopt*[c], *J.C. Dargon*[b], *V. Keim*[a]
[a]Med. Klinik für Gastroenterologie Mannheim, Universität Heidelberg, FRG, [b]INSERM U315, Marseille, France; [c]Chirurgische Klinik der Universität Tübingen, FRG

We recently described a rapid and efficient technique for the separation of rat pancreatic secretory proteins by FPLC [Fuchs et al., J. Chromatogr. 1992;576:297–304]. We are now reporting a method to separate human pancreatic juice and to purify the human pancreatitis-associated protein (PAP).

Methods. Human pancreatic juice was collected from patients with combined renal and pancreatic transplantation. In a first step separation was achieved by a Mono-S column and a mixed salt- and pH-gradient (10 m*M* LiCl to 2 *M* LiCl and pH 5.3 to pH 6.3). After that a phenyl-superose with a decreasing ammonium-sulfate gradient was used to purify PAP.

Results. Using human pancreatic juice up to 10 well separated peaks were found. In the eluate the main pancreatic secretory proteins were detected by their enzymatic activities, and identified by their molecular weights and isoelectric points. No autoactivation of the proteins occurred during separation. Human PAP could be identified in one peak after the mono-S run, which was contaminated mainly with proelastase. This fraction was applied to a phenyl-superose column. Pure human PAP could be eluated with 1 *M* $(NH_4)_2SO_4$. SDS- and 2-D-gels showed a homogenous PAP peak. After transfer to nitrocellulose the purified product was detected with antibodies to human PAP.

Conclusion. The FPLC technique enabled the separation of human pancreatic juice without activation of enzymes. Pure human PAP could be isolated from the juice using a cation-exchange and a hydrophobic-interaction column.

42 On the Mechanism of Pancreatic Hyperplasia Induced by Profound Duodenogastric Reflux

T. Gasslander, H. Mukaida, M.K. Herrington, R.A. Hinder, T.E. Adrian
Departments of Biomedical Sciences and Surgery, Creighton University School of Medicine, Omaha, NE, USA

An increased incidence of pancreatic cancer has been reported in patients with previous gastric surgery. We have previously shown that pronounced duodenogastric reflux (DGR), induced by a split, gastro-jejunostomy in rats, causes pancreatic growth (hyperplasia). This trophic effect is accompanied by an increase in circulating cholecystokinin (CCK) concentrations. Because duodenal contents pass through the stomach in this model, the resulting acidification may inactivate trypsin and result in CCK elevation through disruption of the feedback loop.

We studied the mechanism of DGR-induced pancreatic growth by inhibition of acid secretion (omeprazole, 40 μmol/kg/day, n = 7), gastrin-receptor blockade (L365,260, 200 nmol/kg/h, n = 8), and CCK-receptor blockade (MK329, 25 nmol/kg/h, n = 9) compared to DGR alone (n = 9). The DGR groups were also compared to sham operated animals with small bowel transection.

Pancreatic weight was significantly increased by DGR ($p<0.001$) and this effect was reduced by all three agents (all $p<0.05$). None of the drugs, however, completely abolished the growth response. Treatment with omeprazole or MK329 had no effect on pancreatic weight in sham operated controls.

These findings suggest that both gastrin and CCK may be involved in DGR-induced pancreatic growth. Furthermore, inactivation of trypsin by gastric acid may be responsible for the elevation of CCK. Hormonal growth stimulation may be involved in the reported higher incidence of pancreatic cancer following gastric surgery and associated DGR.

43 Relationship between CCK and PP Responses to a Meal and Exocrine Pancreatic Function in Healthy Subjects and in Chronic Pancreatitis

B. Glasbrenner, E. Dominguez-Munoz, O. Pieramico, M. Büchler[a]*, R. Riepl*[b]*, P. Malfertheiner*
Departments of Internal Medicine and [a]General Surgery, University of Ulm, and [b]Medizinische Klinik Innenstadt, University of Munich, FRG

Postprandial exocrine secretion is stimulated by cholecystokinin (CCK) and inhibited by pancreatic polypeptide (PP). The aim of this study was to investigate CCK and PP responses to a meal in relation to exocrine pancreatic function (EPF) in healthy subjects and patients with chronic pancreatitis (CP).

Patients and Methods. 14 patients with different stages of CP (mean 41 years, range 24–56) and 14 healthy subjects (mean 39 years, range 22–53) were studied. CP was staged by ERCP and CT. EPF was assessed in all 28 subjects using the pancreolauryl serum test (PLT). For evaluation of endogenous hormone release, an oral test meal of 1430 kJ (32 kJ% fat, 24 kJ% proteins, 44 kJ% carbohydrates) was consumed. Plasma samples were taken for 30 min before (–30, –15, 0 min) and 5, 10, 20, 30, 40, 50, 60, 90 and 120 min after test meal. CCK and PP plasma levels were measured by highly specific RIAs. The mean of the 3 preprandial samples was taken as basal plasma level. Integrated hormonal responses were quantified by the area under the curve over the whole study period.

Results. Basal CCK and PP plasma levels were not different between patients with CP and controls, and were not correlated. However, a direct linear correlation between CCK and PP responses to the meal was found in healthy subjects ($r=0.74$, $p<0.005$). This coupling of postprandial hormonal responses was not evident in patients with CP ($r=0.16$; n.s.). In patients with CP and steatorrhea (SCP) (n = 6), peak fluorescein serum concentration was <2.5 μg/ml, and CCK and PP responses to the meal were significantly impaired (CCK response = 61.3 ± 13.69 in SCP vs. 109.8 ± 14.03 in controls, $p<0.05$; PP response = 3920 ± 1773 in SCP vs. 13418 ± 3299 in controls, $p<0.05$). In CP patients with mild/moderate exocrine insufficiency (PLT>2.5 μg/ml), CCK and PP responses varied greatly. However, PP (but not CCK) response was significantly correlated with the PLT result in this subgroup of CP ($r=0.70$; $p<0.01$).

Conclusion. Our data point to a physiological interrelationship between CCK and PP responses to a meal. This link is interrupted in patients with CP and exocrine insufficiency.

44 Peptide YY (PYY) – a Physiological Regulator of Endogenously Stimulated Pancreatic Secretion in Man?

D. Grandt, P. Layer, K. Rippel, M. Schimiczek, C. Kölbel, O. Ahrens, C. Beglinger[a]*, J.R. Reeve Jr.*[b]*, V.E. Eysselein*[c]*, H. Goebell*
Departments of Gastroenterology, Universities of Essen, FRG, [a]Basel, Switzerland, [b]Harbor UCLA, and [c]VA Wadsworth, USA

It has been shown that PYY inhibits gastric acid secretion and exocrine pancreatic secretion in dog, rat and pig. In man, however, PYY inhibits gastric acid secretion, but not exogenously, i.e. caerulein and secretin stimulated exocrine pancreatic secretion. Therefore the physiological importance of PYY for regulation of exocrine pancreatic secretion in man is unknown. It was our aim to determine whether PYY may contribute to the regulation of human exocrine pancreatic secretion.

Methods. We studied the effect of graded doses of PYY on endogenously stimulated pancreatic enzyme secretion in 8 healthy subjects. After an overnight fast subjects were incubated with a multilumen duodenal tube. A mixture of essential amino acids (450 μmol/min) and marker solution (PEG, 45 mg/min) was perfused into the duodenum to induce moderate endogenous stimulation of

Results. (Table to Abstract 44)

PYY dose, pmol/kg/h	Trypsin, U/min	Chymotrypsin, U/min	Bile acids, umol/min	Plasma PYY, pmol/l
0 (NaCl)	81.3±28.4	37.2±10.3	9.4±2.1	8.1±1.0
15	50.3±11.8*	22.6±4.6*	7.8±2.0	46.8±9.0*
45	33.4±10.2*	20.4±3.9*	7.4±1.6	80.9±18.0*
135	27.1±7.3*	16.6±4.1*	7.1±2.2	177.6±52.3*

Mean values±SEM; *p<0.01 vs. NaCl, n=8.

pancreatic secretion. Peptide YY (PYY 1-36) was infused intravenously, and duodenal trypsin, chymotrypsin and bile acids outputs were measured.

Conclusion. PYY infused in a dose mimicking physiological postgrandial plasma levels (15 pmol/kg/h) significantly inhibits pancreatic enzyme but not bile secretion in man. These data suggest that PYY may contribute to the late postprandial inhibition of exocrine pancreatic secretion in man.

45 Peptide YY (PYY) Inhibits Exocrine Pancreatic Secretion in Isolated Perfused Rat Pancreas

D. Grandt[a], *J.R. Reeve Jr,*[c], *C. Eberhard*[b], *O. AlTai*[b], *V.E. Eysselein*[d], *H. Goebell*[a], *M.K. Müller*[b]

[a]Department of Gastroenterology, University of Essen, [b]Klinikum Mannheim, FRG; [c]CURE VA Wadsworth, [d]Harbor UCLA, USA

PYY inhibits centrally stimulated exocrine pancreatic secretion in rat more potently than hormonally stimulated secretion [Putnam et al., Am J. Phys. 1989]. PYY does not inhibit amylase output in isolated acini. Therefore a central site of action for PYY has been assumed. However, a peripheral site of action on nerve fibers or blood vessels cannot be excluded. We therefore studied the effect of PYY and its receptor subtype specific analogues Pro34 PYY and PYY 3-36, which selectively bind to Y1 and Y2 receptors, respectively, on the hormonally stimulated amylase secretion in extrinsic denervated rat pancreas.

Results.

% of amylase output of controls SD		n
Controls	100±13	6
PYY 1-36	46*±6	6
Pro34 PYY	54*±9	6
PYY 3-36	65*±7	6

* p<0.05.

Methods. We used an isolated perfused rat pancreas preparation as previously described. Exocrine secretion was stimulated by CCK-8 (100 pg/ml; 12 ml/h). Peptides under study were constantly infused at 2 nmol/l (4 ml/min) and amylase secretion was measured.

Discussion. Y1 and Y2 receptor agonists of PYY inhibit exocrine pancreatic secretion in the extrinsically denervated pancreas model. The localization of these receptors is uncertain, but may be on blood vessels (Y1 receptors) or on intrinsic pancreatic nerve fibers. These findings demonstrate an intrapancreatic site of action for PYY, although the physiological importance of this mechanism remains unclear.

46 Does PYY Inhibit Canine Exocrine Pancreatic Secretion by Activation of Y2 Receptors?

D. Grandt, S. Teyssen, M. Schimiczek, E. Niebergall, H. Goebell, M.V. Singer, J.R. Reeve Jr., V.E. Eysselein

Universities of Essen and Heidelberg, FRG, Harbor UCLA, and CURE VA Wadsworth, USA

PYY and NPY share common binding sites with similar affinity. Two receptor subtypes, Y_1 and Y_2, have been characterized, but their importance for inhibition of exocrine pancreatic secretion in dog is unknown. We studied the effects of a new Y_1 selective agonist (Pro^{34}PYY 1-36) and a Y_2 selective agonist (PYY 3-36) in comparison to the effects of PYY 1-36 and NPY 1-36 on exocrine pancreatic secretion in dog.

Methods. We prepared 8 dogs with gastric and duodenal Thomas-like cannulas. Dose response studies of pancreatic bicarbonate and protein outputs to IV administration of PYY and NPY, given against an IV background of secretin (20.5 pmol/kg/h) and caerulein (29.6 pmol/kg/h), were done. After 1 h background infusion, PYY 1-36, PYY 3-36, NPY 1-36 and Pro^{34}PYY were administered at two consecutive doses of 200 and 400 pmol/kg/h. In control experiments IV infusion of NaCl was used. T test was used for data analysis (*=p<0.05).

Results. (Table to Abstract 46)

pmol/kg/h	Protein (mg/15 min, SEM)			Bicarbonate (μmol/15 min, SEM)		
	0	200	400	0	200	400
PYY 1-36	263 (51)	183 (39)*	174 (24)*	2,001 (489)	1,211 (379)*	922 (351)*
NPY 1-36	306 (81)	271 (92)	281 (84)	2,284 (612)	1,969 (603)	2,041 (579)
PYY 3-36	259 (53)	207 (41)*	179 (34)*	2,289 (460)	1,650 (360)*	1,198 (328)*
Pro^{34} PYY	305 (88)	254 (85)	304 (113)	2,300 (601)	2,065 (529)	2,030 (395)
NaCl 0.9%	210 (48)	257 (56)	215 (40)	2,116 (456)	2,440 (417)	1,992 (304)

* $p<0.05$.

Conclusions. Pancreatic protein and bicarbonate secretions were inhibited by PYY 1-36 (Y_1&Y_2 agonist) and PYY 3-36 (Y_2 agonist). The Y_1 agonist Pro^{34}PYY (even at 2000 pmol/kg/h, data not shown) did not inhibit exocrine pancreatic secretion. Thus inhibition of hormonally stimulated exocrine pancreatic secretion seems to be Y_2 receptor mediated. The Y_2 receptors involved are preferring PYY, since in the doses tested PYY 1-36 but not NPY 1-36 inhibited secretion.

47 Overexpression of a 89-kDa Heat Shock Protein in Human Pancreatic Adenocarcinoma

T. Gress[a], F. Müller-Pillasch[a], C. Weber[a], U. Lacher[a], H. Frieß[b], M. Büchler[b], G. Adler[a]
Departments of [a]Internal Medicine I and [b]Surgery, University of Ulm, FRG

Heat-shock induced proteins (hsps) have been shown to elicit tumour-specific immunity to a number of tumours [Shrivastava et al., Sem. Immunol. 1991;3:57–64], with hsps belonging to the 90 kDa group being of particular relevance. In this project we performed transcription studies to examine the involvement of hsps in human pancreatic adenocarcinoma. RNA was extracted from human pancreatic tissues (adenocarcinoma, chronic pancreatitis and normal control) and transferred to Northern blots. Northern blots were hybridized with labelled recombinant DNA probes containing sequences coding for a 89-kDa heat shock protein (several other hsps are being tested in parallel) to assess differences in the transcript levels. Labelled RNA probes from the same hsps were used for in situ mRNA hybridization with histological sections of paraffin-embedded pancreatic tissue (control, carcinoma, chronic pancreatitis). Hsp 89 was significantly overexpressed in adenocarcinoma as compared to chronic pancreatitis and normal pancreatic tissue. No significant differences were found to occur between chronic pancreatitis and normal pancreas. Thus, the increase in transcript levels of hsp 89 in pancreatic adenocarcinoma seems likely to be directly involved with malignant transformation, rather than being a nonspecific response of pancreatic cells to chronic injury. We conclude that hsp 89 is consistently overexpressed in human pancreatic carcinoma, suggesting a functional involvement of hsps in the carcinogenesis in the pancreas (e.g. tumour-specific immunogenicity, antigen processing and presentation).

48 Enhanced Ileal Lipase Outputs by Protease Inhibition in Pancreatin-Treated Patients with Pancreatic Insufficiency (PI)

L. Guarner, M. Antolin, L. Chicharro, J.-R. Malagelada
Digestive System Research Unit, Hospital General Vall d'Hebron, Barcelona, Spain

We have previously shown that treatment of severe PI with enteric-coated pancreatin increases ileal outputs of both trypsin and lipase, but steatorrhea is only partially corrected. We reasoned that partial inhibition of luminal proteolytic activity might protect lipase and further increase its activity. Thus, we expanded upon our previous study [GE A187, 1989] by an additional group of 8 patients with advanced PI. All were intubated with a 230-cm multilumen tube that allowed simultaneous double marker perfusion at two sites (a 20-cm duodenal segment with PEG and a 20-cm ileal segment with phenol red). Luminal pH, trypsin and lipase outputs were quantified at 30-min intervals for 4 h in each patient after ingestion of a standard liquid meal. There were in total 8 patients on placebo, 8 patients who received 5 capsules of enteric coated (pH 6 release) pancreatin (40000 lipase FIP and 2250 trypsin FIP total) and 6 other patients who received similar enzyme treatment plus a protease inhibitor incorporated into the capsules.

Duodenal lipase and trypsin outputs were similar in both groups receiving enzymes. By contrast, ileal lipase outputs were higher in patients receiving protease inhibitor than in those who did not (log median 3.3, range 2.8–5.1, vs. 3.1, range 0–4.2, respectively). Differences were significant ($p < 0.05$) vs. placebo but did not reach statistical significance between both enzyme-treated groups. Median ileal trypsin outputs showed a trend towards being lower in protease-inhibitor treated patients (3.4 vs. 3.7). Duodenal pHs were similar in all three groups (5.9, 5.2–6.2 for placebo, 5.8, 4.9–6.4, for pancreatin alone and 6.0, 5.2–6.2, for pancreatin plus protease inhibitor). These preliminary results suggest that 'in vivo' partial inhibition of luminal proteases by a protease inhibitor added to pancreatin improves exogenous lipase delivery to the distal small bowel in patients with severe PI.

49 Non-Operative Management of Necrosis in Acute Pancreatitis

S. Guillaumes, I. Blanco, P. Clavé, E. Piedrafita, L. Marruecos, C. Peréz, A. Farré, F. Lluís
Lab. Investig. Gastrointestinal, Servicios de Cirugía, Patología Digestiva, Radiodiagnóstico y Cuidados Intensivos, Hospital de Sant Pau, Barcelona, Spain

The management of necrosis in acute pancreatitis (AP) ranges from the modality of non-operative treatment to the early and extensive resection of necrotic tissue (necrosectomy).

Aim. Evaluate the results of non-operative management of patients with necrosis and severe AP.

Patients and Methods. A total of 326 patients with AP (gallstones 62%, alcoholic 16%) were studied over 38 months. Necrosis was assessed by enhanced CT scan in patients with severe AP. CT-guided fine needle aspiration was used to detect infection. Surgical or percutaneous drainage was applied only in the presence of infection.

Results. Necrosis was found in 44 patients (13.5%): by CT in 37, at surgery in 6, and by ultrasound in 1 patient. Among patients with necrosis, final diagnosis of infection was obtained in 17 of them (38%): by fine needle aspiration in 9, by CT in 4, and at surgery in 4. Infection was diagnosed during the first two weeks of disease in 5 patients; in 12 during or after the third week. The likelihood of infection increased with the extension of pancreatic and peripancreatic involvement (from 18% for CT scan grade C to 50% for CT grades D–E). Overall mortality of 3% in the series increased to 15.9% in the group of patients with necrosis. Infected necrosis resulted in a 29.4% mortality rate, whereas the mortality of non-infected necrosis was only 7.4%. Mortality rate of 60% in the presence of early infection (i.e. during the first 2 weeks) decreased to 16.6% when infection was detected after the second week.

Conclusion. In our series, non-operative management of non-infected necrosis results in a low mortality rate. In severe AP infection appears to be the leading cause of death.

50 Effect of Chloroquine on Activity of Lysosomal Enzymes in Pancreas of Mice with Acute Pancreatitis

S. Guillaumes, I. Blanco, A. Villanueva, P. Clavé, D. Sans, A. Chabás, A. Farré, F. Lluis
Lab. Investig. Gastrointestinal, Hospital de Sant Pau, and Institut de Bioquímica Flor de Maig, Barcelona, Spain

In a previous study [Digestion 1991;49:22–23] we have shown that chloroquine (CQ), an inhibitor of lysosomal enzyme activity, increases survival of mice with CDE diet-induced acute pancreatitis (AP).

Results. (Table to Abstract 50)

	nmol/mg/prot		Latency (%)	
	vehicle	CQ	vehicle	CQ
Normal pancreas				
Cathepsin B1	209 ± 74	76 ± 10*	53 ± 19	47 ± 6
Acid phosphatase	478 ± 118	236 ± 39*	33 ± 12	35 ± 13
β-hexosaminidase	279 ± 35	149 ± 37*	54 ± 9	28 ± 22
β-glucuronidase	90 ± 28	32 ± 7*	34 ± 17	42 ± 13
Acute pancreatitis				
Cathepsin B1	309 ± 39**	269 ± 20	9 ± 1	35 ± 6*
Acid phosphatase	506 ± 33	451 ± 74	8 ± 4	48 ± 11*
β-hexosaminidase	187 ± 28	137 ± 30*	15 ± 5	50 ± 9*
β-glucuronidase	52 ± 8	33 ± 4	7 ± 4	30 ± 6*

Mean ± SD; *$p < 0.05$ vs. vehicle; **$p < 0.05$ vs. normal pancreas.

Aim. Determine the effect of CQ on activity of the lysosomal enzymes cathepsin B1, acid phosphatase, β-hexosaminidase, and β-glucuronidase in the pancreas of normal mice and mice with CDE-induced AP.

Material and Methods. After a 24-h fast, female CD1 mice (3–4 weeks old, BW 12–16 g; n = 6/group) were given either normal diet or CDE diet. Mice were then fasted for 24 additional h. Vehicle or CQ-diphosphate (50 mg/kg BW) was given by intraperitoneal route at 0, 24 and 48 h. Mice were sacrificed at 56 h, the pancreas was removed and homogenized. Enyzme activity was measured fluorometrically using methyl-umbelliferyl substrates. Latent activity (latency), a measure of lysosomal stability, was determined for each enzyme.

Summary. In normal pancreas, CQ inhibited the activity of all enzymes studied without altering their latency. In pancreas with AP, treatment with CQ did not modify the activity of cathepsin B1 and acid phosphatase, whereas it increased the latency of all enzymes. Cathepsin B activity increased in pancreas with AP.

Conclusion. The beneficial effect of CQ on experimental AP appears to be independent of cathepsin B1 and may be mediated by its effect on lysosomal stability.

51 Chronic Treatment with Chloroquine Results in Subcellular Redistribution of Lysosomal Enzyme Activity Similar to That Found in Acute Pancreatitis

S. Guillaumes, A. Villanueva, I. Blanco, P. Clavé, D. Sans, A. Chabás, A. Farré, F. Lluis
Laboratorio Investigación Gastrointestinal, Hospital de Sant Pau, and Institut de Bioquímica Flor de Maig, Barcelona, Spain

Development of autophagic vacuoles that result in subcellular redistribution of lysosomal enzyme activity is a common finding in several experimental models of acute pancreatitis (AP). Chloroquine (CQ), a lysosomal enzyme inhibitor, induces autophagic vacuoles in several tissues.

Aim. To determine the effect of chronic treatment with CQ on subcellular distribution of the lysosomal marker acid phosphatase in normal pancreas.

Material and Methods. Experiment 1: female CD1 mice (n = 6/group) were given either normal diet or CDE diet and were sacrificed at 56 h. Experiment 2: CD1 mice (n = 6/group) were given vehicle or CQ (50 mg/kg BW, bid) by intraperitoneal route for 10 days, and mice were then sacrificed. Pancreas was removed and homogenized. Subcellular fractionation was performed using differential centrifugation, and the following enriched fractions were obtained: zymogen granules, lysosomal, microsomal, and soluble. Acid phosphatase activity was measured fluorometrically using a methylumbelliferyl substrate.

Summary. Redistribution of the lysosomal enzyme marker acid phosphatase from the lysosome-enriched fraction to the zymogen granule-enriched fraction was similar in both experiements.

Conclusion. Our findings suggest that this lysosomal phenomena may be an unspecific pathway in acinar pancreatic cell.

Results. (Table to Abstract 51)

	Experiment 1		Experiment 2	
	normal diet	CDE diet	vehicle	CQ
Homogenate	439 ± 111	539 ± 19	275 ± 47	149 ± 34*
Zymogen granules	293 ± 110	821 ± 121*	241 ± 44	420 ± 69*
Lysosomal	732 ± 244	641 ± 85	882 ± 28	507 ± 94*
Microsomal	466 ± 122	577 ± 34	405 ± 102	642 ± 52*
Soluble	281 ± 84	445 ± 77*	376 ± 98	171 ± 12*

Mean ± SD; *$p < 0.05$ vs. normal diet or vehicle.

52 Intra-Duodenal Fatty Acids and Not Triglycerides are Responsible for the Release of CCK in Human

R. Guimbaud, J. Moreau, M. Bouisson, J. Escourrou, N. Vaysse
Gastro-Entérologie et INSERM U151, CHU Rangueil, Toulouse, France

A failure of CCK release after meal has been shown in patients with exocrine pancreatic insufficiency (PI). This abnormality could be restored by the administration of pancreatic extracts [Gut 1989;30:866] or gastric lipase [Gastroenterology 1991;100:A267], suggesting that digested rather than intact nutriments are responsible for the release of CCK and subsequently gallbladder contraction in man. The aim of this study was to determine the specific role of different lipidic stimuli in patients with PI.

Methods. Seven male patients (mean age 52 years) with PI secondary to chronic pancreatitis were selected. PI was considered severe in 5 of them (lipase output $<1{,}000$ UI/min) and moderate in other 2 (lipase output $>1{,}000$ and $<2{,}300$ UI/min). Plasma CCK (by bioassay), gallbladder contraction (by ultrasound) and enzyme outputs (amylase and chymotrypsin) in response to intra-duodenal instillation of either oleic acid as free fatty acids (FFA) or Intralipid® 20% as triglycerides were measured in each patient with at least 48 h interval between each test.

Results. In patients with severe PI, intra-duodenal instillation of FFA generates a more pronounced (91 ± 11 pM vs. 49 ± 21 pM) and faster (15 min vs. 30 min) ($p<0.05$) CCK release than triglycerides. Furthermore, gallbladder contraction was more efficient when FFA were administered in the duodenum instead of triglycerides ($86 \pm 5\%$ vs. $69 \pm 4\%$) at 10 min ($p<0.05$) and ($73 \pm 8\%$ vs. $51 \pm 5\%$) at 15 min ($p<0.03$). Among patients with measurable residual pancreatic function, enzyme outputs were shown to be higher during FFA than triglycerides instillation.

Conclusion. In human (as previously described in dog), FFA rather than triglycerides, when present in the duodenum, stimulate CCK release and gallbladder contraction. In patients with moderate PI this phenomenon may increase residual enzymatic secretion. These results allow to consider the development of enzymatic preparations that cause a fast release of FFA in the duodenum as gastric lipase.

53 Pseudolymphoma of the Pancreas Mimicking Cancer

E. Hatzitheoklitos, M. Büchler, T. Imaizumi, W. Mohr, P. Di Sebastiano, H.G. Beger
Departments of Surgery and Pathology, University of Ulm, FRG; 1st Department of Surgery, Hippokration Hospital, Thessaloniki, Greece

Pseudolymphoma has been described as occurring in a wide variety of sites including the lung, small intestine, stomach and gallbladder. We present a case of pseudolymphoma of the pancreatic body and tail. A 68-year old woman presented with weight loss, recurrent abdominal pain, accompanied by an elevation of serum pancreatic enzymes. Ultrasonographically there was a suspicion of pancreatic tumor in the body and tail. In CT scanning there was an enlargement of the pancreatic body with obstructed pancreatic duct accompanied by dilatation of the pancreatic duct in the tail region. ERCP showed normal duct in the pancreatic head. In the pancreatic body there was a filiform stenosis followed by massive dilatation of the Wirsung duct within the pancreatic tail. Preoperative diagnosis was tumor localized in the pancreatic body with obstructive lesion in the pancreatic tail. During surgical exploration pancreatic body and tail were altered macroscopically comparable with a suspicion for malignancy. A pancreatic left resection was carried out. Histology revealed massive follicular lymphatic hyperplasia based on the presence of hyperplastic follicles with germinal centre and mixed infiltration of plasma cells and mature lymphocytes with no significant cytologic atypia. This is the 2nd case of pseudolymphoma of the pancreas in world literature [Hum Pathol 1991;22:724–726]. Pseudolymphoma of the pancreas seems to be a benign lesion which develops on the basis of chronic inflammation. Occurring at other GI locations in moderate frequency it obviously represents a rare entity in the pancreas.

54 Ontogeny of EGF and TGFα in the Neonatal Guinea Pig Pancreas and the Effects of Exogenous EGF on These Growth Factors

M.K. Herrington, T.E. Adrian
Department of Biomedical Sciences, Creighton University School of Medicine, Omaha, NE, USA

Epidermal growth factor (EGF) and transforming growth factor α (TGFα) have been shown to stimulate cellular proliferation in many tissues in vivo and in vitro. The aim of the current study was to investigate the ontogeny of EGF and TGFα in the pancreas of the neonatal guinea pig, and to observe the effect of exogenous EGF on pancreatic content of these growth factors.

Pancreatic EGF and TGFα were measured at 1, 4, 8, and 14 days of age in 28 untreated neonatal guinea pigs and in 8 adults. In the experimental group, EGF and TGFα were measured after 4 and 15 days of continuous subcutaneous infusion of EGF (30 μg/kg/day), using osmotic minipumps inserted at birth ($n=13$).

Pancreatic EGF content at birth was 13-fold greater than in adults (12.8 ± 1.8 vs. 1.0 ± 0.1 pmol/g, $p<0.001$). Values declined during neonatal life (4 days 7.4 ± 1.6, 8 days 4.6 ± 1.4, 15 days 5.6 ± 1.2). TGFα was also markedly elevated at birth compared to the adult level (1.38 ± 0.16 vs. 0.24 ± 0.04 pmol/g, $p<0.0001$) and declined to adult values by 15 days. EGF infusion resulted in a 64% reduction of EGF content at 4 days and a 68% decrease at 15 days (both $p<0.02$). TGFα concentrations were significantly reduced 4 days (39%, $p<0.001$), but not at 15 days.

The high levels of EGF and TGFα in the neonatal guinea pig pancreas suggest that these peptides, which act through the same receptor, may have an important role in neonatal pancreatic development. Suppression of endogenous EGF and TGFα by infusion of exogenous EGF provides evidence of feedback regulation, further emphasizing the possible importance of these regulatory peptides in the neonatal pancreas.

55 Effect of High Fat Diet and Cholecystokinin Receptor Antagonist on Induction of Pancreatic Tumors in the Hamster

M.K. Herrington, J. Permert, K. Kazakoff, P.M. Pour, T.E. Adrian
Department of Biomedical Sciences, Creighton University School of Medicine, and the Eppley Cancer Institute, Omaha, NE, USA

Considerable etiological evidence links fat intake with pancreatic cancer. A possible mechanism for this is through the hormone cholecystokinin (CCK). Our aims were to investigate the potentiation by high fat diet on pancreatic cancer in a model resembling the human disease, and to investigate the effect of CCK receptor blockade during tumor initiation on cancer incidence.

Four groups of 25 hamsters were studied. At 9 weeks of age, 2 groups of animals received continuous infusion of CCK receptor antagonist (MK329, 25 nmol/kg/h) for 14 days; controls received placebo. Four days after surgery, one MK329 and one control group were fed a high fat semi-synthetic diet (35% fat); the other groups were given a similar composition with low fat content (5%). Four days after these diets were started, all animals received a single injection of N-nitrosobis(2-oxopropyl)amine (BOP, 20 mg/kg BW). Surviving animals were killed and autopsied after 55 weeks. Histology was performed in blind fashion on 3 step sections through the entire pancreas.

The total incidence of malignant tumors was higher in high fat diet animals (49%) than low fat (17%, $p<0.01$). The incidence of advanced lesions (adenocarcinomas) was higher in the high fat (26%) than the low fat group (5%, $p<0.01$). No differences in total malignant lesions or incidence of advanced tumors were seen between MK329-treated animals and controls.

These findings confirm an increased incidence of malignant tumors in animals receiving a high fat diet. The lack of effect of MK329 on tumor incidence provides no evidence for an important role for endogenous CCK during tumor initiation, although this hormone may be involved in tumor promotion by high fat diet.

56 Pancreatic Secretagogues that Activate Phospholipase C Alter Internalization of Endothelins (ET) in Pancreatic Acini by a Protein Kinase C Mechanism

P. Hildebrand, R.T. Jensen
Digestive Diseases Branch, NIDDK, National Institutes of Health, Bethesda, MD, USA

Recently we have shown that rat pancreatic acinar cells possess ET_A and ET_B receptors. In this study we investigated the ability of pancreatic secretagogues to regulate these ET receptors. Agents that stimulate phospholipase C significantly reduced binding of ^{125}I-ET-1 and ^{125}I-ET-3 to acini: 1 μm CCK-8 (75% reduction) > carbachol (30%) > bombesin (Bn) (25%). VIP and secretin which increase cAMP were without effect. The phorbol ester TPA (1 μm) reduced ^{125}I-ET-1 and ^{125}I-ET-3 binding, whereas the Ca^{2+} ionophore A23187 (1 μ*M*) alone had no effect and did not potentiate the inhibitory effect of TPA. The CCK_A receptor antogonist L-364,718 alone had no effect on ^{125}I-ET-1 or ^{125}I-ET-3 binding, but was able to antagonize the inhibitory effect of CCK-8. The ability of CCK-8 to inhibit binding of ^{125}I-ET-1 was dose-dependent, detectable inhibition was seen with 1 n*M* CCK-8, half-maximal with 6.5 n*M*, and maximal inhibition with 1 μ*M*. TPA caused a detectable inhibition at 1 n*M* and 25% inhibition at 1 μ*M* CCK-8. Whereas CCK-8 reduced binding of radiolabeled ETs at 37 °C, it had no effect at 4 °C. The reversibility of ^{125}I-ET-1 or ^{125}I-ET-3 binding was slow and incomplete compared to ^{125}I-(Tyr^4)Bn, suggesting internalization. Removal of surface-bound ^{125}I-ET-1 by base stripping showed that by 30 min 75% and by 120 min 90% of the cell associated radioactivity was internalized. The addition of 1 μ*M* CCK-8 caused a 45% decrease of internalized ^{125}I-ET-1, but did not change the amount bound to the plasma membrane. The protein kinase C inhibitor staurosporine dose-dependently reversed the inhibitory effect of CCK-8 and TPA on ^{125}I-ET-1 binding. Staurosporine (0.3 μ*M*) caused half-maximal reversal of the inhibiting effect of 1 μ*M* CCK-8 on ^{125}I-ET-1 binding. ^{125}I-ET-1 binding to pancreatic acinar membranes was not affected by either 1 μ*M* CCK-8, 1 μ*M* Bn, 1 m*M* carbachol, 1 μ*M* TPA, 0.1 μ*M* VIP or 0.1 m*M* secretin. In conclusion, pancreatic secretagogues that increase phospholipase C inhibit binding of radiolabeled ETs to rat pancreatic acini, whereas agents that act through cAMP do not. Protein kinase C is likely involved in mediating this effect, whereas Ca^{2+} mobilization seems not to be involved. CCK-8 acts through its own receptor and decreases internalization of ^{125}I-ET-1.

57 Nitric Oxide Is an Important Element in Neurally Induced Pancreatic Exocrine Secretion

J.J. Holst, T.N. Rasmussen, P. Schmidt
Department of Physiology C, University of Copenhagen, Denmark

The formation of nitric oxide (NO), enzymatically calatyzed by NO synthases in both endothelial cells and in autonomic nerves seems to explain some non-cholinergic, non-adrenergic tissue reactions. We studied the possible role of NO in vagally induced pancreatic exocrine secretion using isolated perfused porcine pancreas ($n=6$) with intact vagus nerve supply. Electrical stimulation of the vagus nerves (VS, 8 Hz, 10 mA) and infusions of VIP, 10^{-9} *M* were carried out before and after addition to the perfusate of the NO synthase inhibitor Nω-nitro-*L*-arginine methyl ester at 10^{-4} *M* with and without addition of *L*-arginine, 10^{-3} *M*. We also studied the effects of sodium nitroprusside, 10^{-4} *M*.

Results. In all experiments VS and VIP caused a profuse exocrine secretion (44±11 and 44±17 times basal secretion). The inhibitor increased vascular resistance 2-fold, but had little effect on the vascular relaxation caused by VIP and VS. The exocrine response to VS was reduced to 19±5% and the response to VIP reduced to 54±12%. Protein and bicarbonate outputs largely paralleled flow rate. Addition of *L*-Arg to the perfusate restored the responses to VS (to 100±21% of controls) and increased the VIP responses (to 65±11%). Sodium nitroprusside increased rate of secretion 55±19 times over basal and decreased vascular resistance. We conclude that formation of NO is an essential link in VS-induced pancreatic secretion, whereas vasodilatation is less influenced.

58 The Mucinous Pancreatic Ductal Ectasia (MPDE)

C. Iacono[a], *C. Bassi*[a], *G. Zamboni*[b], *S. Corrà*[a], *G. La Guardia*[c], *G.P. Marzoli*[c], *G. Cavallini*[d], *C. Procacci*[e], *P. Pederzoli*[a], *G. Serio*[a]
Departments of [a]Surgery, [b]Pathology, [d]Medicine, and [e]Radiology, University of Verona;
[c]Division of Surgery, Civil Hospital, Bolzano, Italy

The MDPE is a cystic disease of the pancreas described for the first time by Itai in 1986. We report our experience about 9 histologically confirmed cases treated from 1981 to 1992. Our series is composed of 7 men and 2 women, mean age 59.2 (range 41–78). In 6 cases the lesion was located in the head, in 2 cases in the body-tail and in 1 case diffuse to the whole head. The symptoms, that resemble a pancreatitis (dyspepsia, abdominal pain and weight loss) was present for an average of 12.4 months (range 1–84 months). One patient had been treated 4.5 years before with pancreojejunostomy for chronic pancreatitis with ductal enlargement. Plasmatic Ca 19-9 measured in 4 cases was increased in 1 patient only. Eight patients were submitted to US and CT that showed a single or multiloculated cystic formation ranging in size from 2.5 to 8 cm. Endoscopy showed in 2 cases an ulcer of the second portion of the duodenum excreting mucous material, with a normal papilla. In the other 2 cases it showed excretion of mucous material from the papilla. ECRP was carried out in 6 patients and showed ductal modifications (multiple enlargement, lack of filling). In 1 case an intraoperative pancreatography was done showing ectasia in the head and tail and ductal enlargement. The surgical treatment consisted of 6 duodenopancreatectomies, 2 distal pancreatectomies and an internal derivation after biopsy in a 70-year-old patient with tumor involving the whole gland and bad general condition. Of the 8 patients who underwent radical surgery 1 died after 28 months and the others are alive and free of disease after a follow-up period ranging from 4 to 120 months (mean 32.1). The patient who underwent internal derivation is alive after 25 m. In conclusion MPDE is a clinical-pathological entity which differs from other cystic lesions of the pancreas for epidemiological features and for location and relationships with the pancreatic ductal system. On the other hand it presents morphological similarities with the intraductal papillary neoplasm and mucinous cystoadenoma: for that reason resection is mandatory, whenever possible.

59 Positron Emission Tomography – A New Diagnostic Procedure in Discrimination between Chronic Pancreatitis and Cancer of the Pancreas

R. Isenmann, M. Büchler, J. Langhans, H. Frieß, W. Uhl, P. Malfertheiner, M. Ebert, W. Reiche, J. Kunz, M. Müller, S.N. Reske, H.G. Beger
Departments of Surgery, Gastroenterology and Nuclear Medicine, University of Ulm, FRG

Even with ERCP, which is the 'Gold Standard' in diagnosis of cancer of the pancreas, discrimination between pancreatic cancer and chronic pancreatitis may remain a severe diagnostic problem. In this study, we focused on the role of Positron Emission Tomography (PET) in this decision-making process.

Methods. Starting in February 1992, up to now 16 patients with either chronic pancreatitis or cancer of the pancreas received 200–400 MBq of F-18 2-deoxyglucose (FDG) i.v. as a standard preoperative diagnostic procedure. Radiotracer accumulation in the abdominal cavity was recorded with a positron emission tomograph. Enhancement of activity in the pancreatic region was regarded as criterion for pancreatic cancer. All patients underwent surgery and results of histological examination were compared with the preoperative PET results.

Results. Histological examination showed pancreatic cancer in 11 patients and chronic pancreatitis in 5 cases. All 5 patients with chronic pancreatitis showed no glucose accumulation in the pancreatic region. Accumulation was found in 9 patients with pancreatic adenocarcinoma. Only 2 patients failed to have glucose enhancement, although they suffered from histologically proven cancer of the pancreas.

Conclusion. These first data about the value of PET with FDG in differentiation between pancreatic cancer and chronic pancreatitis are promising. PET seems to be a valuable method for establishing the diagnosis of pancreatic cancer. Our data encourage further investigation of this new imaging technique, which also allows the localization of tumor metastasis.

60 Infection of Pancreatic Necrosis – When Does it Happen?

R. Isenmann, M. Büchler, W. Uhl, T. Imaizumi, P. Malfertheiner, U. Pralle, C. Bialas, H. Meyer, H.G. Beger
Departments of General Surgery and Gastroenterology, University of Ulm, FRG

Infection of pancreatis necrosis is an important factor in the course of severe acute pancreatitis. Clinical findings suggest that this happens in the 2nd to 3rd week after onset of the disease. In this study, repeated ultra-sonography guided fine needle aspiration (FNP) was used to determine the time when bacterial contamination occurs.

Methods. Seventeen patients suffering from CT-proven necrotizing pancreatitis (9 male, 8 female, median age 60 years, median Ranson score 3 points, range 1–6), who were admitted to our hospital, entered this study. FNP was carried out at the time of admission and repeated every 3 days. All aspirates were examined bacteriologically in order to reveal infection of necrosis. All patients were treated conservatively until FNP showed bacterial infection.

Results. Two (11.7%) patients died in the latter course of the disease, both of them had infected necrosis. Bacterial infection was shown in 7 patients (41%). In 10 patients, necrosis remained sterile. Bacterial infection could be shown at a median of 6 days after onset of pancreatitis (range 1–25 days). A total of 20 bacterial isolates was detected: 5 strains of Staph. aureus, 5 strains of Enterococci, 3 strains of Pseudonomas, 2 strains of Klebsiella and Streptococcus viridans and 1 strain of Neisseriaceae, E. coli and Candida.

Conclusion. This study supports data which have shown, that bacterial infection of pancreatic necrosis is a frequent finding. According to our results, the crucial time when this occurs is between the 1st and 2nd week after onset of the disease. A repeated FNP seems to be justified in order to select patients with infected necrosis, because this has an important impact on further treatment.

61a Pancreatic Duct Stone ESWL

C. Jakobeit, L. Greiner
Medizinische Klinik A, Klinikum, Wuppertal Barmen, FRG

Introduction. Up to 70% of all patients with chronic pancreatitis have pancreatic duct stones. We have now successfully used shock-waves for treatment of 10 patients with obstructing duct stones.

Methods. All 10 patients (16–59 years) had pancreatic duct stones with a diameter up to 15 mm. All stones could be localized and treated by ultrasound (370–6295 shock-waves, Lithotripter MPL 9000, Dornier, Munich).

Results. All stones could be disintegrated into fragments; in 2 patients the fragments passed the papilla spontaneously, in 6 patients the fragments were extractable endoscopically. In 2 patients extraction of fragments was not possible (stricture of the duct). In 7 patients ESWL led to a significant reduction of 'pancreatic pain'.

61b Effects of Prostaglandins (PG) of E, F and I Series Leukotrienes (LT) and Platelet Activating Factor (PAF) on Amylase Release from Isolated Pancreatic acini

J. Jaworek, S.J. Konturek
Institute of Physiology, University Medical School, Krakow, Poland

Exogenous eicosanoids were reported to affect pancreatic secretion, but it is unknown whether this effect is mediated by the changes in pancreatic circulation or by direct action on pancreatic secretory cells. In this study the effects of PG, LTC4 and PAF and the blockers of their biosynthesis or receptor antagonists on amylase release from the isolated rat pancreatic acini were examined. The acini were incubated with pancreatic secretagogues (caerulein or urecholine) in the presence or absence of various concentrations (10^{-9}–10^{-5} *M*) or PGE2, Nocloprost (stable analog of PGE2), PGI2, PGF1α, LTC4, PAF, indomethacin (inhibitor of endogenous PG formation) and A-63162 (blocker of LT biosynthesis). PGE2, PG1, and PGF1α inhibited secretagogue-stimulated enzyme secretion from isolated pancreatic acini but inhibitory potency was about 1000 times lower on the molar basis, than that of Nocloprost. PAF and LTC4 produced concentration-dependent stimulation of amylase release from the resting acini reaching about 50% of caerulein-induced maximal response and enhanced amylase release induced by caerulein or urecholine. The effects of PAF and LTC4 were reversed by the addition of receptor antagonist for PAF (TCV-309) and for LTC4 (FPL-55712). These results indicate that PG inhibit, while PAF and LTC4 stimulate pancreatic enzyme secretion and thus could be implicated in the control of this secretion under normal and pathological conditions.

62 Histamine H_3-Receptors Inhibit Electrical Stimulation-Evoked Amylase Secretion in the Isolated Guinea-Pig Pancreas

L. Jennings[a], *J. Singh*[a], *R.W. Lea*[a], *J.S. Davison*[b], *K. Sharkey*[b]
Departments of [a]Applied Biology, Lancashire Polytechnic, Preston, UK, and [b]Medical Physiology, University of Calgary, Canada

In a previous study we have demonstrated that histamine can have marked secretagogue effects on enzyme secretion in the guinea-pig pancreas, mediating its responses mainly via activation of histamine H_1-receptors [Pariente et al., Cell Physiol and Biochem 1991;1:111–120]. The present study investigates the possible involvement of histamine H_3-receptors in the control of nerve mediated amylase release in the isolated guinea-pig pancreas. Isolated superfused pancreatic segments were electrically stimulated (EFS) at frequencies of either 10, 20 or 30 Hz (amplitude 20 V, pulse width 1 msec) over a 20-min period. Amylase output was measured using an on-line automated method [Pearson et al., Am J Physiol 1984; 246:563–573].

EPS evoked marked frequency-dependent amylase output from superfused pancreatic segments. Maximal responses were obtained at 20 Hz. When segments were incubated with the histamine H_3-specific agonist R-α-methylhistamine (10 nmol), a significant ($p<0.001$, Student's t test) reduction in amylase output in response to EFS was seen compared to responses obtained in the absence of the H_3 agonist. In addition, using an immunohistochemical technique we have demonstrated the distribution of histamine containing cells primarily around blood vessels in the guinea-pig pancreas. The results of this study indicate that activation of histamine H_3-receptors may play an important negative feedback role in the regulation of neurally mediated pancreatic enzyme secretion in the guinea-pig pancreas.

L. Jennings supported by The Wellcome Trust.

63 Genomic alterations of p53 in Pancreatic Cancer

H. Kalthoff[a], *A. Schmidt*[a], *C. Roeder*[a], *D. Kasche*[a], *G. Lauer*[a], *H. Maacke*[b], *W. Deppert*[b], *K. Pantel*[c], *G. Honold*[c], *E. Scherer*[d], *W. Schmiegel*[a]
[a]Department of Immunology, Medical Clinic, University Hospital, Hamburg; [b]Heinrich Pette Institute, Hamburg; [c]Institute of Immunology, München, FRG; [d]The Netherlands Cancer Institute, Amsterdam, The Netherlands

We have analyzed the expression of p53 in human pancreatic adenocarcinomas by immunostaining, Western blotting, enzyme immunoassay, and by Northern blotting. The data were combined with investigations of genomic mutations in both, the p53 and the c-Ki-ras oncogene. A panel of 14 cell lines reacted positive with the antibodies pab240, pab1801 and CM-1, but not with wild type-specific pab1620. The immunoreactivity observed was confined to the nuclei of the majority of the cells, but some cell lines showed a significant cytoplasmic immunostaining. On tissue cryosections 9/13 tumor specimens revealed a positive staining with CM-1 antibodies, mostly with the nuclei of transformed cells, however, some ductal cells in the tissue also exhibited a profound cytoplasmic staining. Ductal cells of normal pancreatic tissue were not stained by CM-1 antibodies. Overexpression of p53 protein was confirmed by Western blotting in 3/8 carcinoma cell lines and further quantitated by an enzyme immuno-assay which detected more than 0.2 ng p53/mg total protein in 9/14 cell lines. On the genomic level 5 cell lines exhibited the wild type sequence of p53 in exons 5–9, 4 of which with low or no detectable p53 protein. In contrast, the highest p53 protein expressing cells displayed mutations at codon 151, 159, 176, 229, and 280, respectively. Immunoperoxidase staining was used to evaluate cytospin preparations derived from pancreatic juice samples collected by ECRP. Positive p53 reactivity was found throughout all diagnostic groups (adenocarcinoma, acute, or chronic pancreatitis), suggesting stabilization of short-lived p53 protein irrespective of mutations in duct cells derived from pancreatic epithelials under in vivo stress conditions. Mutations in p53 have been previously described to functionally cooperate with mutated ras genes in fully transforming cells. Consequently, we analyzed genomic alterations in the c-Ki-ras gene at codons 12, 13, and 61. Point mutations were revealed by PCR and dot blot hybridization in 11 out of 15 (73%) cell lines tested. These frequent activations of a ras-oncogene in combination with mutations of a tumor suppressor gene are likely to contribute to the highly malignant phenotype of pancreatic carcinomas.

Supported by BMFT 01GA08805/6.

64 Induction of the Fed Pattern of Human Exocrine Pancreatic Secretion by Nutrients: Role of Cholecystokinin and Neurotensin

M. Katschinski[a], *C. Dippel*[a], *M. Reinshagen*[a], *J. Schirra*[a], *R. Arnold*[a], *R. Nustede*[b], *C. Beglinger*[c], *G. Adler*[d]
Departments of Gastroenterology, University Hospitals of [a]Marburg, FRG, [c]Basel, Switzerland, [d]Ulm, FRG, and [b]Department of Surgery, University Hospital of Göttingen, FRG

The aim of this study was to assess the role of CCK and neurotensin in converting the cyclical interdigestive pattern of pancreatic secretion into the non-cyclical fed pattern.

Methods. Six healthy male volunteers were studied on 4 separate days. During each experiment, a mixed liquid meal or solutions of individual nutrients were intraduodenally perfused at 2 ml/min. The mixed meal contained 4.3 g glucose, 2 g fractionated soya oil, and 1.7 g casein hydrolysate per 100 ml and delivered a caloric load of 0.9 kcal/min into the duodenum. The isocaloric and isotonic solutions of individual nutrients contained 44.5 g glucose, 17.8 g soya oil, or 44.5 g hydrolyzed serum bovine albumin per liter and delivered 0.36 kcal/min into the duodenum. Duodenal aspirates and blood samples were collected at regular intervals for determination of pancreatic enzyme outputs and plasma levels of CCK and neurotensin, respectively. Pancreatic secretory pattern during nutrient perfusion was classified as cyclical if (a) there was no consistent stimulation vs. basal state and (b) secretory peaks [Gastroenterology 1992;102:1378–1384] occurred. It was classified as non-cyclical in case of consistent stimulation vs. basal state without secretory peaks.

Results. In each individual, the mixed meal converted the cyclical interdigestive pattern into the non-cyclical fed pattern, whereas none of the 3 individual nutrients abolished the interdigestive pattern. Not only the mixed meal but also lipid and protein perfusion consistently stimulated CCK release. Integrated incremental CCK release amounted to 32.3±9.9 pg/ml · 180 min with the mixed meal, 23.2±6.5 pg/ml · 180 min with lipid perfusion ($p<0.05$ vs. mixed meal) and 13.4±3.8 pg/ml · 180 min with protein perfusion ($p<0.05$ vs. mixed meal). The carbohydrate solution did not significantly release CCK (3.5±3.1 pg/ml · 180 min). None of the duodenal perfusates raised neurotensin plasma levels.

We conclude that (a) intraduodenal delivery of a mixed meal at 0.9 kcal/min converts the interdigestive pattern of pancreatic secretion, (b) CCK but not neurotensin is involved in converting this pattern in response to low-caloric meals, and (c) a threshold amount of CCK release must be exceeded to convert the secretory pattern.

65 Cyclical Secretion of Pancreas-Specific Protein in Man

M. Katschinski, J. Schirra, R. Arnold, B. Göke
Department of Gastroenterology, University of Marburg, FRG

Pancreas-specific protein (PASP) is a pancreatic cytosolic protein which is currently tested as a marker of acute pancreatitis. However, its biological role is completely unknown. Therefore, the present study investigated in general if PASP is secreted into pancreatic juice and in particular if its secretory pattern runs parallel to the cyclical fluctuations typical of the interdigestive state.

Methods. In each of 8 healthy male volunteers, pancreatic secretion (outputs of bicarbonate, amylase, lipase, trypsin, and chymotrypsin) and antroduodenal motility were measured during one or two complete interdigestive cycles (10-min sampling periods). Concomitantly, plasma levels and duodenal outputs of PSAP were determined by RIA (Biermann, Bad Nauheim) for each sampling period. According to the grade of contractile activity, phase II was subclassified (separately for antrum and duodenum) into subphases of low (a), intermediate (b), and high (c) activity.

Results. Duodenal output of PASP (Table to Abstract 65)

	Phase I	Phase II	Phase III	ANOVA
PASP (μg/10 min)	59.8±7.6	129.1±24.5	163.0±24.5	0.001
Antrum	Subphase IIa	Subphase IIb	Subphase IIc	
PASP (μg/10 min)	72.5±9.3	143.9±34.9	203.5±37.3	0.0007
Duodenum	Subphase IIa	Subphase IIb	Subphase IIc	
PASP (μg/10 min)	72.9±13.1	124.8±23.2	223.4±38.2	0.0002

Mean±SEM.

Results. (Table to Abstract 66)

Parameter	Phase I	Phase II	Phase III	ANOVA
Amylase (U/10 min)	778.6±83.6	1,842.9±239.9	2,533.4±477.3	0.0001
Trypsin (U/10 min)	192.2±24.8	721.0±87.2	1,519.1±228.7	0.00001
Bicarbonate (μmol/10 min)	1,679.8±639.8	1,995.7±798.7	2,809.2±1,091.0	0.02
Gastric acid (μmol/10 min)	860.9±124.8	1,140.2±134.5	1,720.3±249.1	0.00001
PP (pg/ml)	28.0±4.9	40.5±6.6	60.7±8.8	0.0001
Antrum	Subphase IIa	Subphase IIb	Subphase IIc	ANOVA
Amylase (U/10 min)	958.7±164.5	1,796.4±320.8	3,148.9±457.2	0.0001
Trypsin (U/10 min)	357.7±72.3	689.8±111.1	1,319.9±185.9	0.0001
Bicarbonate (μmol/10 min)	1,688.0±672.7	2,114.5±880.2	2,948.8±1,343.6	0.0001
Gastric acid (μmol/10 min)	872.1±129.4	1,123.1±148.3	1,598.1±214.8	0.0001
PP (pg/ml)	26.8±3.6	35.3±3.4	46.6±5.2	0.0001
Duodenum	Subphase IIa	Subphase IIb	Subphase IIc	ANOVA
Amylase (U/10 min)	999.8±166.0	2,078.7±265.0	3,629.4±477.9	0.0001
Trypsin (U/10 min)	381.1±68.1	859.4±136.7	1,519.1±228.7	0.0001
Bicarbonate (μmol/10 min)	1,687.7±674.9	1,973.6±732.2	2,938.3±1,275.6	0.0001
Gastric acid (μmol/10 min)	848.3±108.0	1,214.3±138.6	1,586.1±195.0	0.0001
PP (pg/ml)	32.3±4.3	33.4±3.8	60.8±13.0	0.001

Mean ± SEM.

With all interdigestive samples pooled, PASP output into duodenal juice was correlated with outputs of amylase ($r=0.67$; $p<0.001$), lipase ($r=0.75$; $p<0.001$), trypsin ($r=0.73$; $p<0.001$), and chymotrypsin ($r=0.74$; $p<0.001$). Correlation with bicarbonate outputs into duodenal juice was less strong ($r=0.25$; $p<0.05$). In all subjects, plasma levels of PASP were consistently below the detection limit of 25 ng/ml.

Conclusions. PASP is secreted into the duodenum. During the interdigestive state, the magnitude of this secretion is closely coupled to the grade of antral and duodenal contractile activity peaking in high-grade phase II and phase III. Moreover, secretion of PASP runs parallel to secretion of amylase, lipase, trypsin, and chymotrypsin, i.e. other secretory products of the acinar cell. Interdigestive PASP plasma levels do not significantly oscillate. Fluctuations of cholinergic tone are suggested to govern the oscillating interdigestive antroduodenal motility and pancreatic enzyme secretion. Since PASP secretion follows this common pattern of interdigestive motility and secretion, it is arguably under cholinergic control.

66 Coupling of Interdigestive Gastropancreatic Secretion and Antroduodenal Motility in Man: Vagal-Cholinergic vs. Hormonal Regulation

M. Katschinski[a], *J. Schirra*[a], *G. Dahmen*[a], *C. Beglinger*[b], *H. Koop*[a], *G. Adler*[c]
Departments of Gastroenterology, University Hospitals of [a]Marburg, FRG, [b]Basel, Switzerland, and [c]Ulm, FRG

The aim of the present study was to investigate during interdigestive state (a) the coupling of antroduodenal motility and gastropancreatic secretion and (b) the role of vagal-cholinergic tone and gastrointestinal hormones as regulators of this coupled pattern.

Methods. In each of 14 healthy male volunteers, antroduodenal motility, gastropancreatic secretion (outputs of gastric acid, bicarbonate, amylase, lipase, trypsin, chymotrypsin) and plasma levels of regulatory peptides (gastrin, PP, CCK, secretin) were measured during 1–3 complete interdigestive cycles (10-min sampling periods). According to the grade of antroduodenal contractile activity, phase II was subdivided (separately for antrum and duodenum) into subperiods of low (a), intermediate (b) and high (c) activity.

Plasma levels of gastrin, secretin and CCK did not fluctuate during the interdigestive cycle.

Conclusions. In the interdigestive cycle, there is a parallel oscillation of gastropancreatic secretion and antroduodenal motility. This holds true for fluctuations among phases I, II and III and within phase II. At variance with PP, a humoral marker of vagal-cholinergic tone, the interdigestive plasma levels of the classical hormonal mediators of gastropancreatic secretion (gastrin, CCK, secretin) do not fluctuate in phase with secretion. The interdigestive pattern of coupled motility and secretion is most likely not under hormonal but under vagal-cholinergic control.

67 The Long-Acting Somatostatin Analogue Octreotide Does Not Inhibit Enzyme Secretion from Isolated Rat Exocrine Pancreatic Lobules and Acinar Cells in vitro

T.P. Kemmer[a], *M. Büchler*[b], *H. Frieß*[b], *C. Längle*[a], *P. Malfertheiner*[a]
Departments of [a]Internal Medicine II and [b]Surgery I, University of Ulm, FRG

The new long-actin somatostatin analogue octreotide (Sandostatin, SMS 201-995) was proven to be a potent inhibitor of human exocrine pancreatic secretion [Köhler et al., Pancreas 1986;1:154–159; Kemmer et al., Aliment Pharmacol Ther 1992;6:41–50]. Aim of the present study was to investigate the inhibitory action of octreotide on basal and secretagogue-stimulated enzyme secretion from isolated exocrine pancreatic lobules and acinar cells in rats.

Methods. Isolated lobules from rat exocrine pancreas were prepared according to Scheele and Palade [J Biol Chem 1975;250:2660–2670], and isolated acinar cells were obtained by a collagenase digestion method. Amylase release from lobules and acinar cells was measured during a 40-min incubation period at 37 °C under placebo and octreotide. Amylase activity was determined by the Phadebas amylase test. Cell integrity was estimated by Azur A dye exclusion.

Results. In pancreatic lobules carbachol (CCh, 10^{-5} *M*) increased amylase secretion from 54 ± 6 %min (mean ± SEM, integrated secretory response over 30 min) to 182 ± 17 %min ($p<0.001$, $n=22$), and in pancreatic acinar cells from 106 ± 14 %min to 275 ± 18 %min ($p<0.001$, $n=10$). Octreotide in concentrations from 10^{-10} to 10^{-5} *M* did not change basal secretion from lobules and cells. CCh-secretion from isolated lobules and cells was not altered significantly.

Conclusion. The potent inhibitory actions of octreotide demonstrated in vivo could not be observed under in vitro conditions. This result points to a humoral mediated rather than to a direct inhibitory effect of octreotide on pancreatic acinar cells.

68 Inhibition of Enzyme Secretion from Isolated Human Exocrine Pancreatic Acinar Cells by the Muscarinic M1-Selective Receptor Antagonist Telenzepine

T.P. Kemmer[a], *M. Büchler*[b], *H. Frieß*[b], *J. Kunz*[a], *P. Malfertheiner*[a]
Departments of [a]Internal Medicine II and [b]Surgery I, University of Ulm, FRG

Acute administration of the new M1-selective receptor antagonist telenzepine (4,9-dihydro-3-methyl-4-[(4-methyl-1-piperazinyl)acetyl]-10H-thieno-[3,4-b][1,5]benzodiazepin-10-one) elicited a marked inhibition of human exocrine pancreatic secretion in vivo. In this study telenzepine was investigated for its influence on basal and secretagogue-stimulated enzyme secretion from isolated human exocrine pancreatic acinar cells.

Materials and Methods. Isolated acinar cells from human exocrine pancreas obtained by surgical resection were prepared by a collagenase digestion technique. Amylase release from acinar cells was measured during a 40-min incubation period at 37 °C under basal and carbachol-stimulated conditions with and without telenzepine. Amylase activity was determined by the Phadebas amylase test (Pharmacia, Freiburg, FRG). Cell integrity was estimated by Azur A dye (0.25%; Fluka, Neu-Ulm, FRG) exclusion.

Results. Amylase secretion was increased by 33, 98, and 23% with carbachol in concentrations of 10^{-6}, 10^{-5}, and 10^{-4} *M*, respectively. Ceruletide (35×10^{-10} *M*) stimulated amylase release by 340%. Telenzepine in concentrations of 10^{-8}, 10^{-7}, and 10^{-5} *M* inhibited carbachol (10^{-5} *M*)-stimulated secretion by 0, 77, and 100%, respectively. Basal secretion was not changed by telenzepine.

Conclusion. The muscarinic M1-selective receptor antagonist telenzepine inhibits carbachol-stimulated enzyme secretion from isolated human pancreatic acinar cells in a dose-dependent manner. M1 receptors on human pancreatic acinar cells play a major role in the regulation of exocrine pancreatic secretion.

69 Effect of Hyperlipidaemia on Three Different Models of Acute Experimental Pancreatitis (AP)

W. Kimura, Th. Kirchner, D. Heß, J. Mössner
Department of Pathology, Medizinische Poliklinik, University of Würzburg, FRG

Lipolytic enzymes and their substrates are discussed to play a role in the pathogenesis of AP. We now investigated whether lipids alter the course of AP in 3 different rat models.

Methods. (1) Edematous AP was induced by repeated s.c. injection of caerulein; (2) necrotizing AP by retrograde duct injection of taurocholate; (3) pancreatic edema by ligation of (a) bile duct at the liver hilus; (b) common bile/pancreatic duct close to the duodenal wall; (c) a + b. Six hours later, rats were sacrificed and isolated perfused pancreas prepared. The pancreata were perfused with either HEPES/Ringer/HAES alone or in combination with various concentrations of lipids (1–5% wt/vol). The activity of lipase and amylase measurable in the portal venous effluate was regarded as a marker of pancreatic injury.

Results. Both in caerulein and taurocholate AP amylase/lipase activities were significantly higher as compared to controls during 45 min of perfusion. In both models addition of lipids caused a dose-dependent marked elevation of enzymes. Ligation (a) did not cause a significant rise of enzymes in the venous effluate; addition of lipids had no effect. Ligation (b) and (c) caused a significant increase of pancreatic enzymes which was further increased by addition of lipids.

Conclusion. The pancreatic damage caused via 3 different pathogenetic ways, i.e. hyperstimulation, disruption of membranes by detergents, ductal hypertension, could be markedly increased by the addition of lipids.

70 Quantification of Pancreatic Microcirculation in Caerulein-Induced Pancreatitis

E. Klar, W. Schratt, Th. Foitzik, H. Buhr, Ch. Herfarth
Department of General Surgery, University of Heidelberg, FRG

Although caerulein-induced pancreatitis has been characterized in detail by means of light- and electron microscopy very little is known about changes in pancreatic perfusion. Thus, it was the aim of this study to quantify pancreatic microcirculation during caerulein pancreatitis utilizing an established model of intravital microscopy.

Method. Twelve New Zealand white rabbits were anesthetized with pentobarbital i.v. and ketamine i.m. Together with the duodenal loop the pancreas was exposed to allow for continuous intravital microscopic observation including video recording. FITC-dextran 70 was injected i.v. as plasma tracer. Caerulein pancreatitis was induced by i.v. administration of caerulein (25 μg/kg/h) ($n = 6$). Blood cell velocity and the number of perfused capillaries was assessed before starting the caerulein infusion as well as 30 min, 1, 2, and 3 h thereafter. In the control group without pancreatitis ($n = 6$) the quantification of microcirculation was performed at comparable intervals.

Results. In caerulein pancreatitis an increase in permeability was recorded by an extravasation of FITC-dextran within 30 min after i.v. infusion of caerulein was started. The number of perfused capillaries remained constant throughout the observation period. The percentage of capillaries displaying high blood cell velocity (>1,000 μm/sec) increased from 37 ± 4% to 75 ± 11% after 30 min, reaching a maximum of 84 ± 8% at 60 min ($p < 0.05$). In contrast, no change in permeability or blood cell velocity was documented in the animals without pancreatitis.

Conclusion. Caerulein-induced pancreatitis is characterized by an increase in microvessel permeability in the early phase. In parallel, hyperemia develops as documented by a homogeneous capillary perfusion at increasing blood cell velocity.

71 Release of Nitric Oxide Upon Vagal and Meal-Induced Stimulation of Exocrine Pancreatic Secretion in Dogs

S.J. Konturek, J. Bilski, W.W. Pawlik, K. Czarnobilski, P. Gustaw, R. Sendur
Institute of Physiology, Medicine School, University of Krakow, Poland

Pancreatic secretion in response to a meal is usually assumed to be mediated by complex neurohormonal mechanisms including local noncholinergic nonadrenergic (NANC) neurotransmitters. Recently, nitric oxide (NO) has been proposed as NANC neurotransmitter involved in the control of gastric motility, microcirculation and secretion. This study was designed to determine the role of NO in the mechanism of exocrine pancreatic secretion in dogs. In conscious dogs with chronic pancreatic fistulas, sham-feeding (SF) and ordinary meat feeding resulted in the increase in pancreatic protein secretion reaching about 32 and 62% of the maximal response to exogenous CCK (400 pmol/kg/h). Pretreatment with *L*-arginine (*L*-Arg) (50 mg/kg/h), a substrate for NO synthase, almost doubled the basal pancreatic secretion and prolonged the peak secretory response to SF or feeding. Similar effects were observed following infusion of nitroglycerin (0.25–4 mg/kg/h). After administration of L-NNA (2.5 mg/kg/h), an inhibitor of NO syntethase, both the SF- and feeding-induced pancreatic protein secretion were markedly inhibited by about 85 and 80%, respectively. When *L*-Arg was combined with L-NNA, the reduction in pancreatic secretion was significantly attenuated. In anesthetized dogs, the *L*-Arg and nitroglycerin caused an increase in the pancreatic blood flow as measured by laser Doppler, whereas *L*-NNA resulted in a marked reduction (by about 30%) in this flow and in secretory response to secretin. We concluded that NO plays a major role in the vagally stimulated and postprandial pancreatic secretion and this is probably mediated by a marked enhancement in pancreatic blood flow.

72 Distribution of Trypsin(ogen)s in Human Pancreatic Acinar Cells: An Immuno-Electron Microscopic Study

B. Krüger[a], *W. Tessenow*[a], *W. Dummler*[a], *M. Büsing*[b], *U. Hopt*[b], *G. Adler*[c]
[a]Institute of Pathobiochemistry, University of Rostock; [b]Department of General Surgery, University of Tübingen; [c]Department of Internal Medicine I, University of Ulm, FRG

Distribution of trypsin(ogen)s in acinar cells of a healthy pancreas and of 11 grafted pancreata (9 biopsies before and 16 biopsies after reperfusion) has been studied by immunogold labelling, electron microscopy and morphometric evaluation. Total trypsin(ogen) (Tn) was marked by a single gold label with a non-differentiating polyclonal antibody. Anionic (aTn) and cationic trysin(ogen)s (cTn) were marked specifically by double labelling with a monoclonal and a polyclonal antibody, and 5 nm and 15 nm gold, respectively.

In zymogen granula of pancreatic grafts especially prior to reperfusion we observed a significantly higher density of Tn labelling in comparison with that of the control pancreas. Additionally, 20 and 50 min after reperfusion more frequent secretory vacuoles of varying electron density and size were labelled for Tn. Five to 6 h after reperfusion large autophagosomes, which labelled homogeneously for Tn, appear. Disintegration of secretory vacuoles results in a free distribution of Tn over the cytosol of several acinar cells. Moreover, the cTn/aTn labelling ratio of 17 graft biopsies, evaluated up to now, significantly decreased (medians: 0.10 and 0.11, before and after reperfusion, respectively) in comparison with that of the control pancreas (median 0.45; $\alpha<0.001$). These changes originate from a significant increase ($\alpha<0.001$) of the labelling density for aTn in granula of the grafted pancreata: 80 μm^{-2} and 57 μm^{-2}, prior to and after reperfusion, respectively, vs. 20 μm^{-2} in the control pancreas (median values in each case).

Our immunocytochemical findings, which are similar to those found after experimental pancreatitis in rats, support the conception of the so-called graft-pancreatitis after pancreas transplantation in man.

73 Blood Glucose Modulates Exocrine Pancreatic Secretion in Man

W.F. Lam, A.A.M. Masclee, A.M. van der Zon, C.B.H.W. Lamers
Department of Gastroenterology and Hepatology, University Hospital of Leiden, The Netherlands

In patients with diabetes mellitus a reduction in exocrine pancreatic function has been observed. Since gastrointestinal motility is dependent on serum glucose concentrations we have investigated whether hyperglycemia influences exocrine pancreatic secretion. Seven healthy volunteers (2 men, 5 women; age 20–26 years) were studied twice, in random order during normoglycemia (5 mmol/h) and during hyperglycemia (15 mmol/h) using a glucose clamp technique. Duodenal juice was collected in 15-min periods through a triple lumen tube using PEG as a recovery marker. Pancreatic and biliary secretions were measured unstimulated for 60 min and for 60 min during infusion of CCK 0.5 IDU/kg/h.

Hyperglycemia significantly reduced basal pancreatic and biliary secretion. CCK significantly increased pancreatic enzyme and biliary secretion both during eu- and hyperglycemia. HCO_3 was not stimulated by CCK and remained significantly reduced during hyperglycemia. CCK-induced pancreatic trypsin output remained significantly reduced during hyperglycemia. It is concluded that exocrine pancreatic secretion is influenced by blood glucose concentrations.

Results. (Table to Abstract 73)

Output/h	Basal		CCK	
	5 mmol/l	15 mmol/l	5 mmol/l	15 mmol/l
Bilirubin (μmol/h)	17 ±5	2 ±1*	77 ±13**	50 ±7**
HCO_3 (mmol/h)	2.6±0.5	1.3±0.1*	2.3±0.3	1.5±0.1*
Amylase (KU/h)	8 ±2	2 ±1*	14 ±2**	10 ±2**
Trypsin (U/h)	22 ±4	4 ±1*	61 ±10**	44 ±9*, **

*p<0.05 compared to 5 mmol/l; **p<0.05 compared to basal.

74 The Clinical Course of Acute Pancreatitis Does Not Parallel Morphological (CT) Findings

P.G. Lankisch, M. Haseloff, R. Becher
Departments of Internal Medicine and Radiology, Municipal Hospital of Lüneburg, FRG

This study compared the clinical course of acute pancreatitis (AP) with its morphological findings as shown upon computed tomography (CT).

Eighty-three consecutive patients with AP were studied (44 men, 39 women). Etiologies were alcoholism in 35%, biliary tract disease in 39% and unknown in 22%. In 3 patients, AP occurred postoperatively and in 1 following ERCP. A contrast-enhanced CT was performed within 72 h after admission, before the patient was discharged, and 3 months later (n=27). Staging of CT findings was performed according to Balthazar et al. [1985, 1990].

Eight patients (10%) died, the remaining patients were discharged and were free of pain 3 months later. They showed normal CRP as well as amylase and lipase levels. The following table shows the CT findings:

Score	On admission	On discharge	3 months later
0	6%	19%	74%
1–3	58%	53%	26%
4–6	33%	24%	–
7–10	4%	4%	–

Thus, CT findings showed abnormalities in 81% of patients upon discharge and in 26% upon follow-up examination.

Conclusion. The clinical course of AP does not run parallel to its morphological findings (as shown by CT). Apparently, the gland takes more time to return to normal after an acute attack.

75 The Delay in Diagnosing Chronic Pancreatitis

P.G. Lankisch, M. Peiper, A. Happe-Löhr, J. Otto, F. Seidensticker, F. Stöckmann
Departments of Internal Medicine, Municipal Hospital of Lüneburg, and of Gastroenterology and Endocrinology, University of Göttingen, FRG

The time interval between the onset of symptoms and diagnosis is well known for malignant but not for benign gastrointestinal diseases. This study aimed at determining the length of delay in diagnosing chronic pancreatitis (CP).

In 327 CP patients (257 men, 70 women), whose diagnoses had been made from 1965–1992, the time interval was estimated between the onset of abdominal complaints (patients with painless CP were excluded) and diagnosis (abnormal secretin-pancreozymin test). CP was alcohol-induced in 72% and idiopathic in 25% of patients. Five had hereditary CP and 2 primary hyperthyroidism. Pancreatic calcifications and pseudocysts were present in 36 and 26%, respectively.

The mean diagnostic interval was 63 months, approximately the same in men and women (64 vs. 58 months). Diagnosis was made earlier in alcoholics than in non-alcoholics (55 vs. 75 months). The time interval was longer in patients with calcifications than in those without (73 vs. 57 months), but shorter when pseudocysts were present (53 vs. 66 months).

Conclusion. More thorough medical instruction is necessary so that CP may be considered as a possible cause behind upper abdominal pain of otherwise unclear etiology. Early recognition and consequent treatment may help avoid complications.

76 Is Lipase/Amylase Ratio Helpful in the Early Etiological Differentiation of Acute Pancreatitis?

P.G. Lankisch, M. Petersen
Department of Internal Medicine, Municipal Hospital of Lüneburg, FRG

An early etiological differentiation of acute pancreatitis may be helpful in deciding whether endoscopic diagnostic/therapeutic procedures are necessary. A lipase/amylase ratio >2 has been thought to specifically indicate alcoholic acute pancreatitis [Gastroenterology 1991;101:1361].

The lipase/amylase ratio was estimated in 87 consecutive patients with acute pancreatitis within the first 24 h after admission. Thirty patients suffered from alcoholic, 38 from biliary acute pancreatitis. In 3 patients, acute pancreatitis occurred postoperatively and in 1 following ERCP. In the remaining 16, etiology was unknown.

A lipase/amylase ratio >2 was found in 73% of alcoholics and in 57% of the patients with non-alcoholic pancreatitis. Sensitivity of the ratio was 73%, specificity 43%, positive predictive value 37%, and negative predictive value 78%. When a ratio of >2.5 was taken for differentiation, sensitivity and specificity increased to 68 and 70%, and the positive and negative predictive values were 54 and 80%, respectively.

Conclusion. A lipase/amylase ratio of >2 or >2.5 may be used to exclude though not prove alcoholism as the etiology of acute pancreatitis.

(Table to Abstract 77)

Trial group	Fatal attacks	Complicated attacks	MOSF incidence	APACHE-II (median/iqr)		
				0 h	24 h	48 h
SCM	19 (17.8%)	28 (26.1%)	15 (14.8%)	8 (6–12)	9 (6–12)	8 (5–12)
IPT/SCM	19 (19.8%)	23 (24.0%)	18 (18.8%)	9 (6–11)	9 (5–11)	8 (5–11)

77 A Prospective, Multi-Centre, Controlled Clinical Trial of Intra-Peritoneal Trasylol Therapy for Acute Pancreatitis

M. Larvin[a], *C. Wilson*[b], *D. Heath*[b], *D. Alexander*[a], *M.J. McMahon*[a], *C.W. Imrie*[b]
[a]Academic Surgical Unit, The General Infirmary, Leeds; [b]Department of Surgery, The Royal Infirmary, Glasgow, Scotland, UK

Fatal acute pancreatitis (AP) is usually associated with multiple organ system failure (MOSF). MOSF can be initiated by agents generated early in the course of AP because of uninhibited intra-peritoneal protease activity. Saline lavage removes the peritoneal exudate, but does not improve outcome, probably because of fluid sequestration, respiratory embarassment and albumin loss. Adding an anti-protease such as aprotinin (Trasylol®, Bayer) may improve efficiency whilst reducing lavage volume. Pharmacodymamic studies demonstrated an effective regimen, instilling 5×10^5 k.i.u. intra-peritoneal Trasylol (IPT) in 1000 ml 0.9% saline, repeated 8 h later. The aim of the study, in which 23 UK hospitals co-operated, was to determine whether IPT could prevent or ameliorate early MOSF, and thus improve the overall outcome of AP.

710 patients with AP (clinical diagnosis, hyperamylasaemia >1,200 IU/l) were assessed (369 men, 341 women; median 57 years, range 14–97; gallstones 34%, alcohol 19%). Patients were randomised to either standard conservative management (SCM) or SCM plus IPT, if aged 16–84 years, history <72 h and no contra-indication to lavage/IPT. Obviously mild AP (no peritonism/hypoxia/hypotension) was excluded. End points were mortality, major complications, MOSF incidence and severity, monitored during the first 7 days using serial APACHE-II scores. Of 208 patients randomised, 3 refused consent, 2 cannulations failed, leaving 107 SCM and 96 IPT+SCM patients, of comparable age, sex, cause, duration of symptoms, Imrie and initial APACHE-II scores.

There were no statistically significant differences between groups. Given the statistical power of the study, it appears that IPT neither prevents or ameliorates early MOSF, nor does it influence the overall outcome of AP.

78 Enzyme Pellet Size and Luminal Nutrient Digestion in Pancreatic Insufficiency

P. Layer, G. Gröger, D. Dicke, M. v.d. Ohe, H. Goebell
Department of Gastroenterology, University of Essen, FRG

Resistant particles are emptied from the stomach simultaneously with meal nutrients only if their median size does not exceed 1.4 mm [GE 1988;94:1315], and there is evidence that with pancreatin microspheres of <1.4 mm, digestion in pancreatic insufficiency is more effective [Z Gastroent 1991;24:417]. To investigate the effects of particle size of pancreatin microspheres on gastric emptying of chyme components and nutrient digestion by direct luminal measurements, 5 patients with severe pancreatic insufficiency (<5% of normal by secretin-cerulein test; no previous abdominal surgery) swallowed a multilumen tube for duodenal and distal-jejunal aspiration and marker perfusion. Each patient received on 3 occasions a nonradioac-

Results. (Table to Abstract 78)

	5×20 mm	1.8–2.0 mm	1.0–1.2 mm
50% Gastric emptying time			
Meal/marker (min)	44 ±2	37 ±7	34 ±10
Lipase (min)	79 ±13[c]	45 ±8[a]	24 ±4[a, b]
Cumulative prandial delivery			
Duodenal lipase (kU)	3.2±1.0	6.1±2.7[a]	6.2±2.5[a]
Jejunal triglyceride (g)	4.2±0.3	2.4±0.6[a]	1.5±0.2[a, b]

Mean values±SE; p<0.05 vs. 5×20 mm; [b]vs. 1.8–2.0 mm; [c]vs. meal.

tive-labelled standard meal (starch, 41 g; triglycerides, 10 g; protein, 11 g) together with a pancreatin capsule (lipase, 10 kU FIP; Kali Chemie, Hannover), either with acid resistant micropellets of 1.0–1.2 mm or 1.8–2.0 mm, or as an acid resistant macrocapsule (5×20 mm), in double-blind randomized order.

Analogous results were obtained for trypsin and protein, and amylase and carbohydrate, respectively.

Summary, Conclusions. In the early (≤30 min) postprandial period duodenal lipase delivery and triglyceride digestion were greater with the smaller, compared with the larger, microspheres; both were equally effective in the later (>30 min) period. By contrast, delivery of lipase from macrocapsules was asynchronous with the meal which explains their low efficacy. Adapting the size of pancreatin micropellets to that required for normal prandial gastric emptying of solid particles markedly improves their digestive efficacy in pancreatic insufficiency.

79 Which Factors Influence the Growth Inhibition by Sandostatin in Metastatic Endocrine Pancreatic Tumors?

H. Lenze, M.E. Trautmann, C. Neuhaus, J. Stapp, C. Bruns, R. Benning, K. Joseph, R. Arnold, and the German Sandostatin Group
Department of Internal Medicine, Division of Gastroenterology, Philipps University of Marburg, FRG

In the German Sandostatin Multicenter Trial an antiproliferative effect of Sandostatin (SAS) therapy has been observed in 13% of the patients with endocrine pancreatic tumors (n=32) including 8 gastrinomas, 16 non-functioning tumors, 1 insulinoma, 3 carcinoid syndromes, and 4 glucagonomas. The aim of this study was to evaluate which factors influence the outcome of SAS treatment. The group of patients responding to SAS therapy with inhibition of tumor growth and the group of non-responders did not differ with regard to age and sex. Neither time elapsed between first tumor symptoms or diagnosis to initiation of treatment were significantly different. Prior therapy with α-interferon or cytostatics did not influence the SAS effect. In general, pancreatic tumors had a worse prognosis than intestinal tumors. Neither suppression of serum hormone levels nor presence of SAS receptors did correlate with the antiproliferative effect of SAS treatment:

Tumor growth inhibition		+	–	Sum
Hormonsuppression	yes	2	1	
	no	5	6	
Receptors +		3	3	6
–		0	1	1

The duration of tumor growth suppression by SAS was limited and variable. The escape from the SAS effect is unlikely due to loss of receptors because we could demonstrate high concentrations of receptors in 2 gastrinomas at the time of escape. In conclusion, the effect of SAS on tumor growth is variable and yet unpredictable.

80 Endoscopic Retrograde Cholangiopancreaticography (ERCP) in Acute Pancreatitis (AP) Reveals Signs of Chronic Disease in Patients with Alcoholic but Not with Other Origin of the Disease

H.-G. Leser, A. Heinisch, H.J. Brambs, B. Bühler, M. Lausen, K. Rückauer, E.H. Farthmann, J. Schölmerich
Departments of Medicine and Surgery, Universities of Freiburg and Regensburg, FRG

In chronic pancreatitis ERCP is the accepted method to define the extent of morphological changes of the pancreatic duct system. It has been suggested that alcoholic AP represents a flare-up of a chronic disorder. In this study morphological alterations detected by ERCP in AP were studied in correlation to the etiology of the disease. ERCP was done in 106 patients with AP as defined by clinical and laboratory parameters during the first 24 h after admission. Interpretation of ERCP was performed according to criteria generally accepted for chronic pancreatitis. Patients could be divided into 3 groups according to the etiology of AP (biliary 39%, alcoholic 22%, undefined 39%). Patients with biliary AP showed pathological changes of their papilla in 40.5%, whereas patients with AP of other origin did so only in 28%. In contrast, alterations of the duodenal mucosa could be detected in 46% of patients with alcoholic AP but only in 13% of patients with AP of other origin. Pathological changes of biliary ducts and gallbladder were seen in 80% of patients with biliary AP, whereas these were found in the remaining patients only in up to 25%. Imaging of the pancreatic duct system revealed pathological changes mainly typical of chronic pancreatitis in 87% of patients with alcoholic AP. Patients with AP of biliary and undefined origin showed only in 47 and 38% alterations of the pancreatic duct system, respectively, and these were mainly related to distal obstruction (i.e. dilatation of the main duct). These findings are in accordance with hypothesis that alcoholic AP occurs in an already morphologically altered organ.

Conclusion. Our data suggest that ERCP in alcoholic AP mainly demonstrates morphological alterations of the duodenum and the pancreatic duct comparable to those seen in chronic pancreatitis, whereas changes of the biliary system and the papilla are mainly found in biliary AP.

81 Dose Response Treatment Studies with Two Pancreatic Carcinoma Cell Lines in vitro

K.H. Link, D. Kindler, M. Hummel, M. Büchler
Department of General Surgery, University of Ulm, FRG

Aim. The increase of regional concentrations might improve the therapeutic efficacy in inoperable pancreatic cancer (PC). To facilitate drug selection for high concentration therapy, available medical treatment modalities should be preevaluated in vitro.

Materials and Methods. Two cloned cell lines were established from a PC-primary tumor [PMH 2/89] and malignant ascites [PMH 3/89]. The cell lines were characterized and showed all signs of human PC. They were used for in vitro toxicity studies in a H^3-thymidin (Thy) soft agar assay according to Kern. Adriamycin (ADM), Cis-platinum (CDDP), 5-FU, mitomycin C (MMC), mitoxantron (NOV), pirarubicin (PIRA), streptocotocin, cortisone (C), cyproterone (C), estrogen, fosfestrol (H), progesterone, tamoxifen, testosteron, and interferon-α (IFN-α) were tested for concentration dependent inhibition of H^3-Thy uptake.

Results. Concentration response of toxicity was observed with ADM, PIRA, CDDP, NOV, MMC, 5-FU, C, H, and IFN-α. Comparing concentration $\times$ time products inhibiting H^3-Thy uptake by $>50\%$, the sequence of cytostatic drug activity was: PIRA, ADM, NOV, CDDP, MMC, 5-FU.

Conclusion. PIRA, ADM, NOV, CDDP, MMC, and 5-FU are suitable to treat PC at high concentrations. H and IFN-α could be effective in PC.

82 Ischemia Reperfusion Syndrome in Experimental Pancreas Transplantation

M.A. López-Boado[a], *A. Vicens-Justo*[a], *G. Hotter*[b], *J. Roselló*[b], *F. Pi-Siqués*[a], *E. Gelpí*[b], *L. Fernández-Cruz*[a]
[a]Department of Surgery, Hospital Clinic, University of Barcelona, and [b]Molecular Pathology Unit, CID-CSIC, Barcelona, Spain

The generation of oxygen free radicals (OFR) is one of several factors that may cause cellular damage after ischemia and reperfusion in transplantation procedures. It is known that lipid hydroperoxides activate phospholipase A_2 (PLA_2) inducing arachidonic acid metabolism.

Aims. To investigate the relationship between arachidonic acid metabolism, PLA_2 activity, and lipoperoxidation induced by OFR after a prolonged period of ischemia in rat pancreas transplantation, and the effect of free radical ablation with superoxide dismutase (SOD).

Material and Methods. Twenty-four rats received pancreas isografts; 8 nonsurgically manipulated rats were used as controls (group I). Isografts were perfused and preserved in Euro-Collin's solution for different periods: group II (n = 8), 45 min cold storage, group III (n = 8), 18 h cold storage, group IV (n = 8), 18 h cold storage plus SOD immediately before reperfusion, Tissue samples were obtained 30 min after reperfusion to determine: thromboxane B_2 (TxB_2) and 6-keto PGF_1, PLA_2 activity and lipoperoxidation by TBA assay.

Results. A period of 18 h of cold preservation induced significant increases of tissue TxB_2 and 6-keto PGF_1 compared with groups I and II, which were concomitant with those observed for PLA_2 activity and lipoperoxides. In all cases SOD attenuated the observed increases.

Conclusion. This results suggest that both activation of PLA_2 and liberation of arachidonic acid metabolites due to the action of OFR may be attenuated with SOD.

83 Camostate-Induced Pancreatic Growth: Importance of Various Intracellular Regulatory Mechanisms of the Polyamine Metabolism

C. Löser, U.R. Fölsch
I. Medical Department, Christian-Albrechts University, Kiel, FRG

Polyamines, which are essential for cellular growth, are highly regulated by various intra- and extracellular mechanisms. This study was designed to evaluate the role of putrescine degradation via diamine oxidase (DAO) to GABA as well as of the interconversion pathway vis oxidation of acetylated polyamines by polyamine oxidase (PAO) during camostate (FOY 305) induced pancreatic growth in rats.

Methods. Male Wistar rats were either treated with (1) the trypsin inhibitor camostate (200 mg/kg orally twice daily), (2) the ornithine decarboxylase (ODC) inhibitor DFMO (2% drinking water plus 3×300 mg/kg i.p.) plus the PAO-inhibitor MDL 72527 (30 mg/kg/day i.p.), (3) the DAO-inhibitor aminoguanidine (AG) (50 mg/kg/day i.p.), (4) camostate plus DFMO plus MDL 72527, (5) camostate plus AG or (6) saline. Seven to 9 animals per group were killed after 12 h, 1 and 5 days.

Results. DAO activity in pancreas is extremely low and only initially increased after camostate treatment. AG does not alter any of the measured parameters except DAO. DFMO plus MDL 72527 highly significantly inhibit the camostate-induced increases in putrescine, ODC and DNA over 5 days, while the initial significant inhibition of pancreatic weight, DNA polymerase, protein and spermidine are absent after 5 days and spermine is even significantly increased. The initial inhibitory effect on the polyamines and the trophic parameters is much more pronounced as recently seen for camostate plus DFMO alone [Löser et al., Digestion 1989;43:98–112].

Conclusions. Degradation of putrescine by DAO, which is essential in gut, is unimportant in the pancreas. Together with the de-novo synthesis via ODC the interconversion pathway is essential for the intracellular regulation of the polyamine pool. (DFG Lö 459/1–10).

84 Neurotensin Fails to Induce Pancreatic Growth and Polyamine Metabolism in Rats

C. Löser[a], *R. Nustede*[b], *H.J. Becker*[b], *H. Köhler*[b], *U.R. Fölsch*[a]
Departments of [a]Internal Medicine, Christian-Albrechts University, Kiel, and [b]Surgery, University of Göttingen, FRG

Aim of the present study was to investigate the influence of i.p. application of neurotensin (NT), which resulted in mildly supraphysiological and even unphysiological high NT plasma levels, on pancreatic growth and polyamine metabolism in rats. Polyamines are known to be essentially involved in cell proliferation and are established as sensitive early indicators of growth.

Methods. (1) Male Wistar rats (n = 5) received 1, 100, and 200 μg NT/kg BW i.p. every 8 h for 8, 16, 24 h and 10 days. (2) 1 and 100 μg NT/kg were given i.p. 8 times per day for 10 days. (3) Male Wistar rats (n = 15) received either 1 μg CCK/kg BW i.p., CCK plus 1 μg NT, CCK plus 10 μg NT or saline every 8 h for 5 days. The following parameters were analyzed: pancreatic weight, DNA, DNA polymerase, protein, amylase, trypsin, ornithine decarboxylase, S-adenosylmethionine decarboxylase, polyamines, pancreatic histology.

Results. Neither 1 μg nor 100 or 200 μg NT resulted in a significant increase of any measured parameter compared to controls in experiments 1 and 2. Administration of CCK alone resulted in a significant increase in pancreatic weight, protein and spermidine concentration, while simultaneous administration of NT had no further increasing effect on any growth parameter or the polyamine metabolism.

Conclusion. Neither alone nor in combination together with CCK, neurotensin exerts any or any further significant effect on pancreatic growth or the parameters of the pancreatic polyamine metabolism (DFG Nu 42/1–2 and Lö 459/1–10).

85 Is Increased Pancreatic Pressure a Cause of Pain in Chronic Pancreatitis (CP)?

G. Manes[a], *M. Büchler*[b], *P. Di Sebastiano*[b], *O. Pieramico*[a], *P. Malfertheiner*[a]
Departments of [a]Internal Medicine II and [b]General Surgery, University of Ulm, FRG

Whether increased pancreatic pressure is a cause of pain in CP is still debated. In the present study, we evaluated the relationship between pain and pancreatic pressure in CP.

Material and Methods. We studied 12 patients with CP (11 men, 1 woman, 49 years, range 42–58) and 5 patients with cancer of the pancreatic tail (4 men, 1 woman, 53 years, range 47–65) as controls. CP was staged on the basis of morphological (ERCP) and functional (serum-pancreolauryl test) criteria, according to current classifications. During the 2 weeks before surgery, patients kept daily records of the intensity of pain on a linear analogue scale. Intraoperative pressure was recorded by a fine needle introduced into the pancreatic tissue and connected to a pressure transducer. A median number of 3 measurements (range 1–6) were performed in each subject. The mean values were used for the statistical evaluation. In controls, pancreatic pressure was determined in macroscopically normal tissue in the head of the pancreas.

Results. Pancreatic pressure was significantly higher in CP than in the control group (29.9 ± 3,1 vs. 7.2 ± 1.1 mmHg, $p < 0.001$). No relationship was found between the pain score and the pancreatic pressure. Pressure was positively correlated with ductal changes but was not correlated with exocrine pancreatic function. Postoperative pancreatic tissue pressure fell by 15.3% in 4 patients with CP in whom pressure assessment was repeated after surgical decompression. Pain was relieved in all CP patients following surgery.

Conclusion. We could not find a relationship between pain and pancreatic pressure in CP. Ductal changes are the most important factor in determining pancreatic pressure, whereas pancreatic function does not seem to exert a major influence.

86 Anti-Polymorphonuclear Cells Globulin Reduces the Flare-Up of Acute Pancreatitis (AP) in Rats

F. Marotta[a], *D.H. Chui*[b]
[a]Liver Unit Crespi, Niguarda Hospital, Milan, Italy; [b]Norman Bethune University, Changchun, China

The xanthine oxidase system has been indicated as one of the sources of oxygen radicals (O_2^-) in the course of

AP. The aim of our study was to evaluate the role played by polymorphonuclear cells (PMNc)-derived O_2^-. Anti-PMN rat antibody (PMN-Ab) was raised in rabbits. One group of rats received an i.p. injection of 15 ml/kg rabbit PMN-Ab 24 h before caerulein-induced AP (7.5 mg/kg/6 h). Rats were sacrificed 3, 6 and 12 h after the onset of AP. Pancreas histology, α-tocopherol, ascorbic acid, glutathione (GSH), superoxide dismutase (SOD) and enzyme levels were done. Hyperamylasemia was unaffected in PMN-Ab group, but histology and dry/wet weight ratio markedly improved ($p<0.05$). Alpha-tocopherol significantly decreased at 3 and 6 h, as compared to control (μg/g protein: 167±11 and 209±19 vs. 287±8, $p<0.01$), whereas only a mild 6 h decrease (μg/g protein: 220±13, $p<0.05$) occurred in PMN-Ab pretreated rats. SOD (mU/mg protein: 24±4, 19±6, 31±9) and GSH (U/g protein: 48±12, 53±6, 66±14) decrease persisted throughout the study ($p<0.01$). In PMN-Ab group SOD (mU/mg protein: 34±5) and GSH (U/g protein: 72±18) decreased at 6 h ($p<0.05$) with a trend to recover at 12 h. Ulike previous ex vivo studies, this research, by showing the partly protective effect of PMN-Ab, points out the role of PMN as a relevant source of O_2^- in the course of AP.

87 Macrophage Oxidative Burst Ability Is Impaired in Chronic Pancreatitis Patients

F. Marotta[a], H. Kimura[b], T. Nakamura[c], K. Hayakawa[a], K. Takebe[c]
Departments of [a]Surgery II, [b]Oral Surgery and [c]Internal Medicine III, University of Hirosaki, Japan

Following our previous work, the aim of the present study was to further define the (m ∅) function in chronic pancreatitis (CP) patients. A series of 15 CP patients aysmptomatic and free from acute attacks for at least 3 months before the study were studied. Twelve had diabetes. As control groups, a series of 20 healthy subjects, 11 chronic renal failure patients, and 10 diabetics age- and sex-matched were studied. After an overnight fast, blood was withdrawn and 1×10^6 m ∅ suspensions were prepared and the viability was each time over 95%. O_2^- generation in m ∅ suspensions was tested with 2-methyl-6-(p-methoxyphenyl)-3,7-dihydro-imidazol(1,2-a)-pyrazin-3-dependent (MCLA) chemiluminescence reader. Light integrated intensity was expressed as count/min and the peak calculated over MCLA baseline. As compared to healthy controls, CP patients showed a significant lower level of O_2^- (15.3±4.3 vs. 27.2±4.7 cpm $\times10^4$, $p<0.01$). Healthy controls showed a quick rise of peak value (208±35 sec) and a slow descent. CP patients showed a slow rise (374±106 sec) and a rapid fall. MCLA value in CP patients was also significantly lower than either chronic renal failure patients (vs. 24.8±4.9, $p<0.01$) and of diabetics (vs. 22.8±3.2, $p<0.05$). From these results it appears that CP patients have indeed an impaired m ∅ function which is unrelated to flare-ups of CP, to diabetes and to aspecific chronic disease status.

88 Islet Cell Tumors at National Cancer Center in Tokyo: ERCP Features in a Rare Pathology in Japan

F. Marotta, H. Tajiri, M. Yoshimori
Department of Internal Medicine, National Cancer Center, Tokyo, Japan

The incidence of islet cell tumors in Japan is reported to be far lower than in other countries. From 1980 to December 1990 8 cases if islet cell tumors were detected at NCC Hospital in Tokyo. Six of them were diagnosed over a total of 2,949 autopsies. During the same period 150 pancreatic carcinomas were found. The age ranged from 25 to 71 years. Four cases are presented: One case was asymptomatic, 4 had hypoglycemic symptoms and 4 vague abdominal pain (2 with relapsing peptic ulcer). On ERCP: One case had a tiny bead-like pattern of the tail of the main pancreatic duct (MPD); 1 showed a stop with spotted pattern in the secondary branches of the uncinate process; 2 had a stop of the MPD in the body. On CT scan the latter 2 cases had a large tumor mass with liver metastasis. Angiography showed hypervascular tumor in these 4 cases. On histology, a common feature of the tumors was a very vascularized interstitium with sinusoid pattern and focal necrosis. Tumors stained for insulin and pancreatic polypeptide in 1 case, glucagon in 1 and insulin in 2. The overall incidence of insulinoma was about 60% of the islet cell tumors registered during the 11-year survey. Unlike the low diagnostic rate reported for ERCP, these data show that an accurate study of the secondary branches of the MPD duct by ERCP may show evidence of subtle anatomical changes in small tumors.

89 Monocyte Activation in Acute Pancreatitis Is Related to the Degree of Systemic Illness

C. McKay, B. Brooks, G. Gallagher, J.N. Baxter, C.W. Imrie
University Department of Surgery, Royal Infirmary, Glasgow, UK

Aim. To investigate the role of monocyte activation in the systemic effects of acute pancreatitis.

Methods. Nine patients with moderate to severe acute pancreatitis had serial measurements of monocyte function during the course of their admission. Daily APACHE-II score was assessed. Peripheral blood mononuclear cells were obtained by centrifugation on Lymphoprep and these were further enriched for monocytes by positive selection in a magnetic cell separation system using conjugated CD14 microbeads. Monocytes were cultured for 24 h in serum-free medium with and without the addition of bacterial lipopolysaccharide (LPS) 200 ng/ml (E. coli-B5). Supernatants were removed and stored at −70 °C for the subsequent estimation of tumour necrosis factor (TNF) by ELISA.

Results. Monocyte purity was $90 \pm 5\%$. No difference in spontaneous TNF production was seen. Initial results suggest increased TNF production in response to LPS stimulation by monocytes isolated from the patients with a significant systemic illness as assessed by the APACHE-II score.

Conclusion. Prolonged monocyte activation occurs in patients with acute pancreatitis. Dysregulation of TNF production by monocytes may play a role in the pathophysiology of the systemic effects of the disease.

90 Nitric Oxide Modulates Pancreatic Response to Caerulein. Histological and Biochemical Improvement in Acute Pancreatitis

X. Molero[a], *A. Salas*[b], *L. Guarner*[a], *J.R. Malagelada*[a]
[a]Digestive System Research Unit, General Hospital Vall d'Hebron, Barcelona, and [b]Department of Pathology, Hospital Mutua de Terrasa, Spain

Caerulein and CCK increase intracellular cGMP concentration. The role of cGMP in the pancreatic response to this peptides has not been thoroughly defined. Guanylate cyclase is activated by nitric oxide (NO) and NO donors like sodium nitroprusside (SNP). L-nitroarginine methyl ester (L-NAME) blocks NO formation from L-arginine. The aim of the present work was to investigate the effects of NO on the pancreatic responses to caerulein. Biliopancreatic duct and jugular vein of anesthetized rats were cannulated for pancreatic secretion recovery and i.v. infusion of drugs. Optimal (0.5 μg/kg) and supramaximal (10 μg/kg) doses of caerulein were tested with or without L-NAME bolus (30 mg/kg). Acute pancreatitis was induced by i.p. caerulein (20 μg/kg) given every hour for 4 h. Prior to caerulein, rats received L-NAME (30 mg/kg), SNP (0.5 mg/kg) or vehicle. Nine hours after the initiation of the experiments, serum amylase and lipase were determined and pancreata removed for histological examination. Edema, inflammation, vacuolization and necrosis were scored blindly.

Results. L-NAME reduced amylase output in response to an optimal dose of caerulein by 57.3% ($p<0.05$) ($n=5$). A supramaximal dose produced a very weak secretory response ($94 \pm 25\%$), but in combination with L-NAME pancreatic secretion was strongly inhibited ($26.5 \pm 6\%$; $p<0.05$ vs. caerulein alone). Intraperitoneal supramaximal doses of caerulein raised serum amylase concentration to 29.4 ± 3.6 U/ml and lipase to 3.6 ± 0.5 U/ml. Treatment with L-NAME achieved a greater amylase concentration (46.8 ± 3.9; $p<0.005$ vs. caerulein alone), whereas SNP reduced serum concentration (17.1 ± 3.6; $p<0.01$) ($n=8$ per group). In addition, SNP also reduced serum lipase concentration (1.3 ± 0.2; $p<0.001$ vs. caerulein alone). SNP was able to reduce, slightly but significantly, histological damage (7.4 ± 0.2 vs. 6.1 ± 0.2; $p<0.05$) ($n=11$). We conclude that NO appears to modulate pancreatic responses to caerulein and that NO donor drugs may improve the course of caerulein-induced acute pancreatitis.

91 A New Method for Monitoring Intracellular pH (pHi) in Isolated Rat Pancreas Using Confocal Laser Scanning Microscopy

F.Ch. Mooren[a], *R. Stoll*[a], *W. Domschke*[a], *J. Gronzweski*[b], *A.W.H. Jans*[b], *R.K.H. Kinne*[b]
[a]Department of Medicine B, University of Münster; [b]Max-Planck-Institut für Systemphysiologie, Dortmund, FRG

To gain further insight into pHi regulation in pancreas under minimal alterations of organ integrity, a new experimental model was developed in which the method of perfusion of the isolated pancreas was combined with confocal laser scanning microscopy.

After removal of the pancreas the splenic segment was perfused via the celiac artery with a dextran (5%) Ringer solution at a constant flow rate of 2 ml/min and fixed in a temperature-controlled plastic chamber at 37°C.

Cells were loaded with a pH-sensitive fluorescent dye (BCECF) by topical application of 10 μmol/l of the membrane-permeant ester BCECF-AM for 30 min. pH measurements were performed with a Biorad MRC-600 confocal microscope equipped with a HeCd-laser (442 nm) and an Ar-laser (490 nm) for dual wavelength excitation. Dye calibration was done separately by measurements of solutions of defined pH with a BCECF-concentration similar to intracellular conditions. Image analysis enabled the detection and comparison of different cell types. Preliminary results showed an intracellular pH in acinar cells of 7.42 ± 0.06 and in endothelial cells of 7.40 ± 0.05. In both cell types, following a CO_2-pulse (20% CO_2) pH decreased to 7.03 ± 0.08.

This new experimental system enables investigations on intracellular pH regulation of different cell types (e.g. acinar, ductal and endothelial cells) under intact tissue conditions with the original cellular architecture and integrity being maintained. Such an approach might be of specific value for further studies on the pathogenesis of acute experimental pancreatitis.

92 Role of Albumin in Treatment of Experimental Acute Pancreatitis (AP)

J. Mössner, W. Kimura, Th. Kirchner
Medizinische Poliklinik, Department of Pathology, University of Würzburg, FRG

Lipolytic enzymes may disrupt cellular membranes via the release of detergents such as fatty acids and lysolecithin. We, therefore, studied the effect of inhibition of lipolytic enzymes under two different experimental conditions in rats.

Methods. (1) Pancreatic damage was induced by sodium-taurocholate injection into the pancreatic duct of isolated rat pancreas and the organ continuously perfused with either a specific lipase inhibitor (tetrahydrolipstatin; THL), a protease inhibitor (gabexate-mesilate; FUT), or albumin. Organ damage was measured by determinations of lipase and amylase in the portal effluate and by microscopy. (2) Necrotizing AP was induced in rats via retrograde taurocholate injection. This was followed by immediate intraparenchymal injection of either FUT, THL, or albumin.

Results. (1) THL and FUT lowered to a small degree the release of pancreatic enzymes into the portal effluate and lowered the degree of lobular necrosis. Albumin, however, was very effective: it lowered significantly the release of pancreatic enzymes into the venous effluate and it reduced the degree of lobular and focal necrosis. (2) Albumin injection into the pancreatic parenchma at multiple sites prevented the development of panlobular necrosis and lowered the degree of extrapancreatic fat necrosis; THL and FUT had only minor effects.

Conclusion. These results strengthen the concept of a role of lipolytic enzymes in the pathogenesis of AP. Albumin, possibly by its ability to bind detergents such as free fatty acids and lysolecithin, may have therapeutical implications.

93 Modulation of Pancreatic Secretion by Capsaicin-Sensitive Neurons in Anaesthetized and Conscious Rats

C. Nagain, N. Gicquel, J. Chariot, A. Tsocas, C. Rozé
INSERM U239, Faculté Bichat, Paris, France

Interdigestive pancreatic secretion is mainly driven by an endogenous cholinergic tone in the rat, whereas the response to intraluminal nutrients mainly relies upon CCK release (1). The present experiments were devised to measure the changes of pancreatic secretion produced by stimulation or destruction of capsaicin-sensitive sensory neurons.

Methods. Acute urethane anaesthetized and semichronic conscious pancreatic fistula rats were used. Capsaicin was administered either as a single acute intraduodenal (i.d.) bolus injection (0.1 to 3 mg/kg, under a volume of 2 ml/kg) to stimulate topically duodenal sensory fibres while measuring pancreatic secretion, or as several high-dose s.c. injections (100 mg/kg), to inactivate sensory fibres, 10 days before measuring pancreatic secretion. Basal interdigestive secretion, or secretion stimulated by an intraduodenal mixed meal of 21 kJ were studied.

Results. (1) In anaesthetized rats, intraduodenal capsaicin produced a dose-related and moderate stimulation of pancreatic secretion, which was not changed by vagotomy, atropine, hexamethonium or the CCK_B antagonist L365260. It was decreased or suppressed by the previous destruction of capsaicin-sensitive fibres and by the CCK_A antagonist L364718. (2) In conscious rats, i.d. capsaicin increased interdigestive spontaneous secretion. The previous destruction of capsaicin-sensitive sensory fibres did not change significantly the spontaneous interdigestive pancreatic secretion, but significantly suppressed the stimulation produced by i.d. capsaicin. Finally, in sensory denervated rats, the pancreatic response to the i.d. meal was unchanged during the first 2 h, and increased during the 2–4 h period following the meal.

Conclusions. (1) Topical i.d. capsaicin induces an increase of pancreatic secretion, that seems mainly peripheral in anaesthetized rats, and involves CCK_A receptors. This might correspond to the local release of neuropeptides by sensory fibres. (2) Sensory denervation by chronic capsaicin increases the late phase of the pancreatic response to meal, which might correspond to the destruction of an inhibitory sensory component.

1 J Chariot et al., Am J Physiol 1990;259:G198–204.

94 Post-Translational Processing of Intrapancreatic Proenkephalin in Chronic Pancreatitis

D.K. Nelson, M. Büchler, D.L. Lucas, P. Di Sebastiano, H. Frieß, P. Malfertheiner
Mayo Clinic, Rochester, Minn., USA, and University of Ulm, FRG

Endogenous opioids, including methionine-enkephalin (ENK), have been localized to acinar and islet structures within the pancreas, where ENK may serve to modulate exocrine and endocrine secretion via neural inhibition. We have shown that the perineural elements of the pancreas are damaged in chronic pancreatitis (CP), ultimately leading to alteration of the neural environment.

Aim. Given the role(s) of ENK in mediating nociception and pancreatic function, it was of interest to examine the status of ENK-containing elements in CP. Specifically, we investigated the post-translational processing of the proenkephalin A precursor, during which encrypted copies of ENK are enzymatically cleaved from this prohormone.

Methods. Tissue from the pancreatic head of 20 patients with alcohol-induced CP was compared with that from 9 organ donors. Concentrations of ENK in acid extracts were determined by specific radioimmunoassay. Additional aliquots of supernatant were assayed after treatment with trypsin and carboxypeptidase B to enzymatically expose ENK encrypted within the precursor molecule (T-ENK). The ratio of T-ENK to ENK within each tissue sample provided an index of the degree of processing which had occurred.

Results. Concentrations of ENK in pancreatic head were increased in CP as compared to controls (2.9 ± 0.4 vs. 0.8 ± 0.3 ng/g, $p < 0.01$), while T-ENK was decreased (104.4 ± 8.0 vs. 159.8 ± 4.9 ng/g, $p < 0.001$). These changes resulted in a 6-fold greater ratio of T-ENK:ENK in controls as compared to CP ($p < 0.001$). Processing parameters were poorly correlated with exocrine function, but were tightly linked to duration and severity of pain.

Conclusions. Post-translational processing of the proenkephalin molecule in pancreatic tissue appears to be altered in CP. Our results suggest that enhanced processing occurs in the disease state, leaving less 'encrypted' ENK to release by subsequent enzymatic cleavage. Possible explanations include irregular activity of endogenous intrapancreatic enzymes in CP, or changes in processing linked to perineural damage. These observations lead to speculation that aberrant processing in the endogenous opioid system may underlie, or respond to, the persistent pain associated with CP.

95 Ultrasound and Endoscopy Guided Percutaneous Cysto-Gastrostomy in the Treatment of Pancreatic Pseudocysts

M. Neskovski, V. Serafimoski, V. Vasilevski, V. Bidikov, T. Pop-Nikolov, B. Docevski[a]
Clinic of Gastroenterology, [a]Clinic of Digestive Surgery, Medical Faculty Skopje, Macedonia

Ultrasound and endoscopy guided cysto-gastrostomy (US-E-CGS) is a nonsurgical alternative for internal drainage of pancreatic pseudocysts. Cysto-gastrostomy is performed with ultrasound (US) and endoscopy guided percutaneous puncture, with a special tubing set, consisting of a double pig tail catheter 8–10 cm in length and 8F, mounted on a needle with a pusher (Hancke set-Cook®). Needle with catheter is percutaneously, transgastrically introduced into the cyst cavity. During endoscopic control, catheter with its curved ends is placed into cyst and stomach lumen, respectively. In contrast to the original technique of S. Hancke, who performes cystogastrostomy in general anesthesia, we perform it with local anesthesia on the place of skin and stomach puncture, and standard endoscopic premedication. Our study includes 12 patients with pancreatic pseudocysts in close relation to the posterior gastric wall. Average size of the cyst was 12 cm (10–15). The catheter was successfully placed in all patients. In 7 patients, cyst disappeared completely the following day. Three patients with small residual septated sac were successfully treated with additional US guided percutaneous external drainage. Other two had obstruction of CG catheter, 10 days after procedure, resulting cyst abscedation. Both were successfully solved with placement of percutaneous external catheter drainage and lavage with antibiotics and saline solution. Pain and oppression relief, and gain in weight were noticed in all patients.

96 Pancreatic and Synovial-Type Phospholipase A2 in Serum in Acute Pancreatitis

T.J. Nevalainen, J.M. Grönroos
Departments of Pathology and Surgery, University of Turku, Finland

Phospholipase A2 (PLA2) is considered to play a key role in the pathology of various inflammatory diseases. It is the rate-limiting enzyme in the synthesis of eicosanoids. In acute pancreatitis, the catalytic activity of PLA2 in serum correlates to the severity of the disease. The cellular source of the catalytically active PLA2 in serum of patients suffering from acute pancreatitis and other diseases is unknown. Specific immunoassays have been developed recently for the measurement of pancreatic group I PLA2 and non-pancreatic, synovial-type group II PLA2. In the present study, the concentrations of group I and group II PLA2s were measured in sera of patients suffering from acute pancreatitis. The catalytic activity of PLA2 correlated highly significantly to the concentration of synovial-type PLA2 but not to the concentration of pancreatic PLA2. The results suggest that pancreatic PLA2 circulates mostly as inactive enzyme protein in sera of patients suffering from acute pancreatitis while synovial-type PLA2 is responsible for the increased catalytic activity of the enzyme and thus might be associated with the pathophysiology of the disease.

97 Prognosis of Acute Pancreatitis (AP): Marked Reduction of Early Mortality

C. Niederau, J.W. Heise, H. Becker
Departments of Medicine and Surgery, University of Düsseldorf, FRG

The present analysis prospectively evaluates prognosis and course of AP in 152 consecutive patients treated in the Departments of Medicine (n = 93) and Surgery (n = 59) of the University of Düsseldorf between 1987 and 1991.

Results. Age and sex distribution was similar for medical vs. surgical patients (51.9 vs. 48.8 years; men/women 1.7 vs. 2.0). In both groups, alcohol was the leading cause (35%) followed by gallstones (28%), idiopathic AP (24%) and ERCP (7%). AP was significantly more severe in surgical vs. medical patients assessed by CT (CT grades II and III: 67.4 vs. 47.8%), by Ranson's score (3.9 ± 0.2 vs. 2.7 ± 0.2) and by mortality (16.9 vs. 4.3%). However, when mortality was correlated to severity there was no major difference between surgical and medical patients (CT grade I: 0%; grade II: 5–9%; grade III: 32–36%). There were 14 deaths in the 152 patients including 10 surgical and 4 medical patients. Only 2/13 deaths occurred during the 1st week and there was no mortality during the initial 24 h. All remaining deaths were seen at late stages (4–12 weeks after admission). The two early deaths were caused by cardiovascular failure. Almost all late deaths were due to multiorgan failure often associated with septic complications. Only 20/59 surgical patients underwent laparotomy (Ranson's score of 5.4 with 40% mortality). Both CT and Ranson's score reliably predicted a complicated course and to a lesser degree also proved useful for prediction of a fatal outcome. The ROC analysis showed that a cutoff point of ≥3 Ranson's signs picked up >90% of deaths or complicated courses. However, the specificity at this cutoff level was only 40 and 60%, respectively. An increase in the cutoff level to ≥4 Ranson's signs predicted >70% of complications with 80% specificity, and a cutoff level of ≥5 Ranson's signs diagnosed >70% of fatalities with 80% specificity. In general, Ranson's score was more helpful to predict complications than fatal outcome. Measurements of other lab markers such as CRP offered only little additional information compared to clinical and CT gradings.

Conclusion. In the literature [Renner et al., Dig Dis Sci 1985;30:1005] and in the own experience (data from 1980–85), 40–70% of deaths occurred during the 1st week after admission. In the present series, only 15% of deaths occurred during this early phase. The great majority of fatalities occurred at late stages due to multiorgan failure and septic complications which as yet cannot appropriately be predicted by initial clinical or CT gradings. The marked decrease in early mortality is likely due to improvements in intensive care medicine.

98 Changes in Distribution of Pancreatic Lysosomal Enzymes in Different Models of Biliary Pancreatitis

C. Niederau, R. Lüthen, J.H. Grendell
Departments of Medicine, University of Düsseldorf, FRG, and University of California, San Francisco, CA, USA

Redistribution of lysosomal enzymes into the zymogen granule-enriched fraction of acinar cells occurs after different forms of pancreatic damage. The resulting colocalization of lysosomal and digestive enzymes was suggested to be a prerequisite for development of pancreatitis. The present study evaluates the distribution of lysosomal enzymes in compartments of acinar cells in different models of rat biliary pancreatitis.

Methods. Moderate edematous pancreatitis was induced by retrograde infusion of NaCl, bile or 5% taurocholate into the pancreatic duct at physiologic pressure (<15 cm H_2O) using a vertical cannula. More pronounced pancreatitis, assessed by morphology and serum amylase,

Results. Cathepsin B in various cell compartments (Table to Abstract 98)

	U/mg DNA	1,300p	12,000p	105,000p	105,000s
Control	0.12	18%	47%	13%	22%
NaCl – low pressure	0.75*	52%*	33%*	6%*	9%*
NaCl – high pressure	0.38*	43%*	29%*	12%	13%*
Bile	0.74*	38%*	37%*	7%*	18%
Taurocholate	0.57*	40%*	35%*	6%*	19%
Ligation	0.38*	53%*	30%*	6%*	8%*

p = Pellet; s = supernatant; total enzyme given in left column; G values indicated in top of table; *$p < 0.05$ by t test when compared to control, SE not shown.

was induced by retrograde infusion of NaCl via a micropump at higher pressure (20–25 mm H_2O) and by duct ligation at the papilla, respectively. Analyzes were done 6 h after induction of pancreatitis; at this early time point morphological damage just starts to develop.

Different forms of pancreatic damage resulting in mild to moderate pancreatitis were associated with marked increases in pancreatic content of lysosomal enzymes. There was a relative redistribution of lysosomal enzymes from the lysosome-12.000 G fraction primarily to the zymogen granule-1.300 G fraction but also to microsomal and soluble fractions. The redistribution did not correlate with the degree and type of pancreatic damage. In despite of such redistribution, the absolute amount of lysosomal enzymes increased in all fractions. Changes in other lysosomal enzymes, e.g. β-galactosidase, were similar to those in cathepsin B.

Conclusion. The findings suggest a secretory blockade of lysosomal enzymes in early stages of pancreatitis similar to that previously shown for digestive enzymes. Redistribution of lysosomal enzymes may be a by-product of secretory blockade. The data also support the idea that there may be a common secretory pathway for digestive and lysosomal types of enzymes in the pancreatic acinar cell.

99 Lazaroids Protect Against Free-Radical Induced Damage to Isolated Rat Pancreatic Acinar Cells

C. Niederau, H.U. Schulz, G. Letko
Department of Medicine, Heinrich Heine University of Düsseldorf, and Department of Surgery, Medical Academy of Magdeburg, FRG

Lipid peroxidation is thought to contribute to cell injury in acute and chronic pancreatitis. Lazaroids (21-aminosteroids without gluco- and mineralocorticoid activity) have recently been shown to protect against oxidative injury in nervous system cells. The present study evaluates the protective potential of the lazaroids U-78518F, U-74500A, and U-74389F (Upjohn) against radical-induced damage to pancreatic acinar cells.

Methods. Freshly isolated rat pancreatic acinar cells were exposed to two types of reaction mixtures known to generate free radicals: (1) a XOD/HX-system consisting of 20 mU/ml xanthine oxidase (XOD), 500 μM hypoxanthine (HX), 50 μM $FeCl_3$ and 75 μM EDTA, a model which generates radicals primarily via H_2O_2; (2) an ADP/Fe^{2+} system consisting of 5.3 m*M* ADP and 300 μ*M* $FeSO_4$, a model which mediates injury primarily by lipid peroxidation. Cell viability was assessed by both cellular uptake of trypan blue (TB) and by release of LDH.

Results. Both the XOD/HX and the ADP/Fe^{2+} systems caused a time-dependent cell damage. As shown previously TB uptake indicated cell injury earlier and more sensitive than LDH release. Results are therefore given for TB uptake; in general the pattern of results was similar for TB uptake and LDH release. In order to standardize radical-induced damage in various experiments with different doses and types of lazaroids, injury was determined in each experiment at the time point at which 50% of cells were damaged in lazaroid-free XOD/HX and ADP/Fe^{2+} solutions, respectively. At doses between 1 and 50 μ*M* lazaroids significantly reduced radical-mediated damage by 25–55% when compared to cells which were exposed to lazaroid-free XOD/HX and ADP/Fe^{2+} solutions, respectively ($p<0.05$ by t test), with maximal protection seen at approximately 10 μ*M*. U-74500A was the most protective substance in the XOD/HX system, whereas U-78518F most effectively prevented injury in the ADP/Fe^{2+} system. In general, protective potencies for the three substances differed only slightly. Lazaroid concentrations <1–0.5 μ*M* were ineffective for all substances, and concentrations >50–100 μ*M* even accelerated acinar cell injury.

Conclusion. The new lazaroid substances showed protective effects against pancreatic cell injury induced by two different in vitro systems set up to generate free radicals and to induce lipid peroxidation. Further studies have to evaluate the protective effects of lazaroids against radical-induced pancreatic injury in vivo in order to assess their potential for clinical therapy.

100 TPS, TPA and CA 19-9 in the Assessment of Pancreatic Cancer

M.P. Panozzo, T. Meggiato, D. Basso[a], P. Fogar, C. Ferrara, M. Plebani[a], P. Scalon, G. Del Favero
Cattedra di Gastroenterologia e [a]Istituto di Medicina di Laboratorio, Università degli Studi di Padova, Italy

TPS is the specific epitope M3 of TPA and has been proposed in monitoring cell multiplication in cancer patients. Aims of the present study were: (1) to assess the diagnostic utility of TPS in comparison with TPA and CA 19-9 in diagnosing pancreatic cancer (PC); (2) to evaluate the influence of jaundice on the results. We studied 19 control subjects (CS), 42 patients with PC, 29 with chronic pancreatitis (CP), 29 with extra-pancreatic malignancies (EPM), 13 with benign liver diseases and 10 with extra-pancreatic benign diseases (EPBD).

TPS was significantly higher in PC and EPM as compared to CS, CP and EPBD ($F=8.01$, $p<0.001$). TPA showed a similar pattern ($F=4.49$, $p<0.001$). CA 19-9 was significantly higher in PC as compared to all the other studied groups ($F=22.77$, $p<0.001$). Sensitivity and specificity of the three markers in diagnosing PC were: 73.8

and 46.9% for TPS; 54.8 and 76.5% for TPA; 92.5 and 69.2% for CA 19-9. PC patients with liver metastases had higher mean values of TPA ($t = 2.39$, $p < 0.05$) and of CA 19-9 ($t = 3.27$, $p < 0.005$) as compared to non-metastatic PC patients. Significant correlations were detected between TPS, TPA and CA 19-9 on the one hand and indices of cytolysis (S-ALT) and cholestasis (S-ALP, S-GGT and S-bilirubin) on the other. The three markers also correlated with serum albumin ($r = -0.222$, $p < 0.01$ for TPA; $r = -0.302$, $p < 0.001$ for TPS; $r = -0.357$, $p < 0.001$ for CA 19-9).

Conclusions. (1) Serum TPS showed a good sensitivity in diagnosing PC, but it lacked specificity; (2) of the three markers studied CA 19-9 was confirmed to possess the best diagnostic utility in assessing PC; (3) the serum levels of CA 19-9 and TPA, but not of TPS, were influenced by the presence of liver metastases; (4) all the three indices were found to be influenced by the presence of liver damage and of jaundice in particular, which may alter their metabolism.

101 Nonsurgical Management of Pancreatic Pseudocysts

Å. Pap, Z. Berger, L. Bálint, Gy. Bodoky
2nd Department Med., St. Imre Hospital, Budapest, Hungary

Nonsurgical healing rate of chronic pseudocysts treated with parenteral nutrition is only 3%, that of acute cysts is about 30%. Our nonsurgical treatment of pancreatic pseudocysts consists of jejunal feeding through an 8F catheter placed into the 2nd loop of jejunum via the working channel of duodenoscope for 2 weeks followed by endoscopic cystogastro- or cystoduodenostomy. Nutrients (Cosilat, EGIS, Budapest; Nutricomp, Braun-Melsungen, Melsungen), containing 80–100 g of protein and 2000–2500 calories/day were given as continuous jejunal perfusion at 2 ml/min. Ultrasonography and/or CT scan was weekly controlled. If pseudocyst persisted, endoscopic drainage was performed after percutaneous fine needle puncture and labelling of cyst with contrast material (Omnipaque). A short 7–8F internal drain assured patency of diathermic hole between the pseudocyst and the gastrointestinal tract for 4 months, then the drain was extracted endoscopically. 9/9 patients with acute and 9/30 patients with chronic cysts improved significantly during the jejunal feeding. Remaining cases were healed after endoscopic drainage except two cases when distance between the pseudocyst and the gastrointestinal wall was too long.

Conclusion. Endoscopic jejunal intubation and feeding followed by endoscopic decompression in patients with persistent pseudocysts can replace surgical modalities.

102 Role of Nitric Oxide in the Regulation of Pancreatic Blood Flow and Secretion

W.W. Pawlik, P. Gustaw, K. Czarnobilski, R. Sendur, T. Pawlik, S.J. Konturek
Institute of Physiology, University Medical School, Krakow, Poland

Nitric oxide (NO) is considered to play an important role in the regulation of the gastrointestinal functions. In the present study we examined whether NO might play a role in the maintenance of pancreatic blood flow and secretory activity. In 15 anesthetized dogs, the superior pancreatico-duodenal artery blood flow (SPBF), and microcirculatory blood flow (PBF) were determined electromagnetically and by laser-Doppler flowmetry. Protein and HCO_3 outputs were determined in the secretion collected from main pancreatic duct. Control SPBF and PBF averaged 38.7 ml/min and 3.5 Volts, respectively. Following i.v. infusion of secretin at dose of 4 U/kg/h there was an increase of SPBF by 39 and of PBF by 57%, respectively. Secretin-induced protein and HCO_3 outputs reached the values of 35 mg/15 min and 0.34 mEq/15 min, respectively. Inhibition of NO synthase by N-nitro-L-arginine (L-NNA) administered i.v. at dose of 25 mg/kg decreased basal SPBF by 26 and PBF by 32%, respectively. Pretreatment with L-NNA inhibited secretin stimulated protein and HCO_3 outputs and decreased SPBF and PBF by 39, 35, 25 and 40%, respectively. L-arginine (100 mg/kg i.v.) reversed all above effects of L-NNA in the pancreas. Our findings indicate that endothelium-derived NO plays a role in the regulation of pancreatic vasculature at rest and it also mediates the vasodilatory and secretory action of secretin in the pancreas.

103 Diabetes in Pancreatic Cancer Is Not Caused by a Reduced Insulin Production

J. Permert, I. Ihse, H. Arnqvist, H. von Schenck, J. Larsson
Departments of Surgery, Medicine and Clinical Chemistry, University Hospital, Linköping, and Department of Surgery, University Hospital, Lund, Sweden

An association between pancreatic cancer and diabetes has previously been reported. Diabetes has been suggested to be a predisposing factor for pancreatic cancer, while other studies have indicated that the altered glucose metabolism is a consequence of the tumor per se. In this study, the incidence of impaired glucose metabolism and insulin secretion were investigated in pancreatic cancer patients.

Methods. 44 patients were included and the incidence of diabetes and impaired glucose tolerance were studied. Unstimulated plasma insulin and C-peptide were analyzed. In 36 of the patients and in 8 healthy controls, the insulin secretion and glucose metabolic capacity and insulin sensitivity were determined during a hyperglycemic clamp and after glucagon stimulation.

Results. 64% of the patients were diabetic, 11% had an impaired glucose tolerance (IGT), whereas 25% had a normal glucose tolerance (NGT). Of the diabetic patients, 39% required insulin treatment (IRD). Fasting insulin and C-peptide levels were significantly increased ($p<0.01$) in the non-insulin requiring diabetics (NIRD) (135 ± 21 resp. 1050 ± 200 pM) compared to the controls (71 ± 4 resp. 500 ± 61 pM). C-peptide levels in IRD (530 ± 90 pM) were similar to controls. During hyperglycemia, insulin secretion remained unchanged in IGT, NIRD and IRD in contrast to patients with NGT and controls which increased markedly. Glucagon stimulated the secretion in all groups. The glucose metabolic capacity and body insulin sensitivity were significantly decreased in pancreatic cancer patients with IGT ($p<0.01$ resp. $p<0.05$), NIRD ($p<0.001$) and IRD ($p<0.001$).

Conclusions. These results demonstrate that pancreatic cancer is associated with a high incidence of impaired glucose metabolism. Furthermore, the data suggest the mechanism to be rather a consequence of a reduced insulin sensitivity than an impaired insulin secretory capacity.

104 Islet Hormone Secretion in Pancreatic Cancer Patients

J. Permert, J. Larsson, H. von Schenck, T.E. Adrian
Departments of Surgery and Clinical Chemistry, University of Linköping, Sweden; Department of Biomedical Sciences, Creighton University, Omaha, NE, USA

Alterations in insulin secretion have been reported in patients with pancreatic cancer. Recent data also indicate changes in the pattern of secretion of other islet hormones. The present study explores the effect of hyperglycemia and glucagon stimulation on secretion of islet hormones in pancreatic cancer.

Methods. C-peptide, islet amyloid polypeptide (IAPP), glucagon, somatostatin, pancreastatin and PP were determined in fasting state in pancreatic cancer patients before ($n=30$) and after ($n=7$) tumor removal by subtotal pancreatectomy and in healthy controls ($n=25$). Preoperatively, islet hormones were studied during a hyperglycemic clamp and after glucagon stimulation in 20 of the patients and in 8 of the controls, and 7 patients were studied postoperatively. The results were evaluated according to the diabetic state of the patients.

Results. Preoperatively, fasting levels of C-peptide, IAPP, and somatostatin were elevated ($p<0.01$) in diabetic patients compared to controls. During hyperglycemia the secretion of C-peptide and IAPP were unchanged in contrast to the controls who increased their secretion ($p<0.01$). No further differences were noted irrespective of diabetic status. After glucagon stimulation, C-peptide and IAPP levels increased in non-diabetics and in controls ($p<0.001$), whereas only C-peptide decreased in non-insulin-requiring patients ($p<0.01$). No further effects were observed after glucagon stimulation. Postoperatively, fasting plasma levels of all hormones returned to normal, except for somatostatin that remained increased ($p<0.05$). The secretory pattern during hyperglycemia and after glucagon stimulation normalized after tumor removal.

Conclusion. The observed changes in islet hormone secretion may contribute to the diabetic state frequently seen in pancreatic cancer. Furthermore, the study highlights the importance of islet preservation with pancreatic tumor resection.

105 Altered Islet Amyloid Polypeptide Storage in Islet Beta Cells in Diabetes Associated with Pancreatic Carcinoma

J. Permert, G.T. Westermark, T. Christmanson, C. Betsholtz, J. Larsson, P. Westermark
Departments of Surgery and Pathology, University of Linköping, and Department of Pathology, University of Uppsala, Sweden

Islet amyloid polypeptide (IAPP) is almost exclusively produced by the islet β-cells. Experimentally, IAPP induces insulin resistance. In type 2 diabetes, IAPP forms amyloid deposits, localized to the islets of Langerhans. High plasma levels of IAPP have been found in diabetes associated with pancreatic adenocarcinoma. The present study was designed to investigate the precise location of the production of IAPP in pancreatic cancer.

Sections of tumor tissue, adjacent pancreas and unaffected pancreas were studied in 15 patients with pancreatic adenocarcinoma in whom pancreatectomy was performed. IAPP and insulin were demonstrated immunohistochemically with specific antisera. IAPP mRNA was visualized by in situ hybridization.

Neither IAPP nor insulin immunoreactivity was demonstrated in tumor tissue. Beta cells in islets of unaffected pancreas tissue exhibited normal IAPP and insulin immunoreactivity. In contrast, islets close to tumor tissue showed a very weak or negative IAPP immunoreactivity in spite of strong reaction for insulin. Irrespective of immunoreactivity, all islets showed a strong labelling with the IAPP probe at the in situ hybridization while no label was seen in tumor cells. The pattern with low or no IAPP immunoreactivity combined with abundance of IAPP mRNA resembles strongly the findings in type 2 diabetes mellitus.

106 Is Islet Amyloid Polypeptide the Cause of Diabetes in Pancreatic Cancer?

J. Permert, P. Westermark, P.M. Pour, J. Larsson, T.E. Adrian
Departments of Surgery and Pathology, University of Linköping, Sweden; Department of Biomedical Science, Creighton University & Eppley Cancer Institute, Omaha, NE, USA

Diabetes occurs in up to 80% of patients with pancreatic cancer. Recent studies suggest diabetes to be a consequence of the tumor rather than a predisposing factor. Pancreatic cancer patients with diabetes are characterized by a marked degree of insulin resistance, which improves after tumor resection. Islet amyloid polypeptide (IAPP) is a recently discovered hormone which is normally co-secreted with insulin from the pancreatic β-cells. Experimentally, IAPP has been shown to cause insulin resistance in vivo and to reduce glycogen synthesis in vitro. The aim of this study was to investigate plasma IAPP levels in relation to diabetic status.

Methods. Plasma IAPP and C-peptide were measured with radioimmunoassay in 30 patients with pancreatic cancer, 46 patients with other malignancies (stomach, colorectal, liver and lung) and in 25 healthy controls. Oral glucose tolerance test was performed in all pancreatic cancer patients.

Results. Fasting IAPP concentrations were grossly elevated in pancreatic cancer patients with both insulin-requiring diabetes (25.0 ± 3.3 vs. control 8.0 ± 1.0 pM, $p<0.001$) and non insulin-requiring diabetes (31.4 ± 3.8 pM, $p<0.001$), but were normal in patients with other malignancies. IAPP/C-peptide ratio was also increased in the pancreatic cancer patients with diabetes. IAPP levels and the IAPP/C-peptide ratio were normalized postoperatively in all pancreatic cancer patients who underwent tumor resection.

Conclusion. Diabetes in pancreatic cancer was associated with grossly elevated plasma IAPP levels. Furthermore, since IAPP has been shown to be anorectic, its overproduction may contribute to the cachectic state of these patients.

107 Behavior of Soluble Interleukin-2 Receptor, Soluble CD4 and Soluble CD8 during the Course of Acute Pancreatitis

R. Pezzilli, P. Billi, E. Beltrandi, L. Gullo, L. Incorvaia, R. Mancini, M. Maldini
Emergency Department, Central Laboratory, St. Orsola-Malpighi Hospital, Bologna, Italy

During activation lymphocytes secrete glycoproteins related to their surface proteins, including soluble forms of the interleukin-2 receptor (sIL-2R) and of the surface proteins CD4 (sCD4) and CD8 (sCD8). We evaluated the release of these mediators in order to assess the activation of the cellular immune system during the course of acute pancreatitis. Thirty-five acute pancreatitis patients (22 M, 13 F, mean age 64 years, range 16–97) were studied. The diagnosis, based on typical abdominal pain and hyperlipasemia, was confirmed by computed tomography and/or ultrasonography in 33 patients, at surgery in 1, and on autopsy in 1. The pancreatitis was of biliary origin in 22 patients, due to alcohol abuse in 8, due to pancreas divisum in 1, due to Type IV hyperlipoproteinemia in 1 and of unknown origin in 3. Twenty-eight patients had edematous pancreatitis, and 7 necrotizing type. In all patients on admission and daily for the following 5 days, sIL-2R, sCD4 and sCD8 were determined on serum by EIA techniques (T Cell Sciences, USA). The normal values (means ± SD), established on a group of 40 healthy subjects, were 262 ± 49 UI/ml for sIL-2R, 305 ± 107 UI/ml for sCD8, and 22 ± 8 UI/ml for sCD4. Serum concentrations (means ± SD) of IL-2R, sCD8 and sCD4 in acute pancreatitis patients studied are reported in the table.

Serum concentrations of sIL-2R and sCD8 were significantly higher in patients with acute pancreatitis than in healthy subjects, whereas sCD4 levels, from the 2nd to the 6th day of observation, were significantly lower in acute pancreatitis patients than in the control group. No differences in serum concentrations of sIL-2R, sCD8 and sCD4 were found between patients with edematous pancreatitis and those with necrotizing type, nor between patients with biliary pancreatitis and those with other forms of the disease. The results indicate that in early phases of acute pancreatitis there is an activation of the cellular suppressor-cytotoxic immune system and an impaired activity of the CD4 T-lymphocytes.

(Table to Abstract 107)

	1 day	2 days	3 days	4 days	5 days	6 days
sIL-2R (UI/ml)	961 ± 779	936 ± 845	930 ± 840	963 ± 923	905 ± 915	891 ± 848
sCD8 (UI/ml)	637 ± 254	680 ± 259	639 ± 275	667 ± 311	608 ± 259	589 ± 286
sCD4 (UI/ml)	20 ± 22	19 ± 14	15 ± 12	14 ± 12	14 ± 12	17 ± 16

108 Effect of the Somatostatin Analogue Octreotide on the Exocrine Function of the Denervated Human Pancreas

F. Pfeffer, M. Büsing, C. Szgraja, U.T. Hopt, H.D. Becker
Department of Surgery, University of Tübingen, FRG

Multiple studies showed that somatostatin (SMS) inhibits the exocrine function of the innervated human pancreas. The exact mechanism however is unclear. There are only few data about the effect of SMS on the denervated human pancreas. Aim of this study was to evaluate the effect of the long-acting SMS-analogue octreotide (Sandostatin®) on the denervated human pancreas.

10 patients after combined pancreas-kidney transplantation were examined. Pancreas secretion was drained by an intraoperatively placed duct catheter. This allowed collection of pure pancreatic juice. Patients with graft pancreatitis were treated by continuous peritoneal lavage. The concentrations of amylase and lipase in serum, pancreatic juice (PJ) and peritoneal dialysate (PD) were determined immediately before SMS therapy (T1), 1–3 days after beginning (T2), during (T3) and after the end of SMS therapy (T4). SMS was administered between 6 and 44 days s.c. in a dosage of 0.1 to 0.3 mg.

In the beginning of SMS therapy, enzyme concentration in PJ was reduced dramatically (T2: amylase 73%, lipase 55%, volume 48%). Later on, however, enzyme concentrations reached pretreatment levels. Total amount of enzymes as well as total volume remained at lower levels. Enzyme concentrations in the dialysate decreased also (T2: amylase 68%, lipase 55%). Subsequently enzyme values remained on low levels. In serum, enzyme concentrations decreased only gradually (amylase T2: 29%, T3: 30%, T4: 41%).

SMS inhibits exogenous pancreatic secretion of the denervated organ. Ductal secretion decreased immediately. The effect on enzyme concentrations, however, disappeared within a few days. This rapid rise of enzyme concentration in PJ to pretreatment levels may be due to tachyphylaxia. Nevertheless, total amount of secreted enzymes remained reduced by 43%. From a clinical point of view it might be of exceptional interest to know that the exudation of pancreatic enzymes out of inflamed pancreatic glands is depressed by SMS, too.

109 Are Interdigestive Intestinal Motility, Pancreatic Secretion and Pancreatic Polypeptide (PP) Release Coordinated Events in Chronic Pancreatitis (CP)?

O. Pieramico, J.E. Dominguez-Munoz, M. Büchler[a], W. Böck, P. Malfertheiner
Departments of Internal Medicine II and [a]General Surgery, University of Ulm, FRG

In the present study, we investigated the relationship among interdigestive pancreatic secretion, intestinal motility, and PP release in CP.

Materials and Methods. 13 healthy subjects (N) (9 M, 4 F, age 35–57) and 9 patients with CP (9 M, age 25–54) were included. The diagnosis of CP was confirmed by CT and/or ERP. Exocrine function was assessed by the fluorescein dilaurate serum test (PLT). Interdigestive antroduodenal motility was examined using a manometric probe with eight ports. A non-absorbable marker, PEG 4000 (15 g/l) was perfused into the proximal duodenum and duodenal juice was recovered from distal duodenum every 15 min. Blood samples were also drawn for measurement of PP (RIA). Interdigestive motility, release of PP, and outputs of trypsin (T), chymotrypsin (Ch) and amylase (Am) were evaluated for at least 4 h.

Results. In CP, outputs of T and Ch were impaired during phase I, II and III of motility ($p<0.01$). However, Am was decreased only during phase I and II ($p<0.01$). The degree of functional impairment correlated with outputs of T ($r=0.69$, $p<0.05$) and Ch ($r=0.78$, $p<0.05$) but not with Am during phase III. The duration of the secretory cycles (90 ± 4 min N vs. 76 ± 3 min in CP, $p<0.05$) and PP cycle (106 ± 7 min N vs. 63 ± 4 min CP, $p<0.01$) was decreased in CP, whereas the length of the migrating motor complex was unchanged (116 ± 19 min N vs. 124 ± 15 min CP, n.s.). In N, peak protease output and PP release always (100%) coordinated with phase III motor activity, whereas in CP this coordination was observed only in 47.3% of phase III ($p<0.05$).

Conclusions. The interdigestive secretion of proteases, but not amylase, correlates with the degree of pancreatic impairment (in the PLT). CP is characterized by an impaired coordination of interdigestive pancreatic secretion, PP release, and intestinal motility.

110 Epidermal Growth Factor Inhibits Cholecystokinin-Stimulated Amylase Secretion from Pancreatic Acinar Cells by Activating Inhibitory Phospholipase C Coupling G-Proteins

A. Piiper, S. Zeuzem, D. Stryjek-Kaminska, J. Stein, W.F. Caspary
Department of Internal Medicine, Division of Gastroenterology, University of Frankfurt/M., FRG

In the exocrine pancreas, calcium-mobilizing hormones initiate enzyme secretion by binding of the hormone to specific receptors. Subsequent activation of phospholipase C (PLC) by the hormone-receptor complex leads to hydrolysis of phosphatidylinositol 4,5-biphosphate into diacylglycerol and inositol 1,4,5-trisphosphate (IP_3). The activation of PLC by the hormone-receptor complex is mediated by heterotrimeric GTP-binding proteins (G-proteins). In this study, we investigated the role and the mechanism of action of epidermal growth factor (EGF) on basal and cholecystokinin (CCK)-induced IP_3-production and amylase release in digitonin-permeabilized isolated rat pancreatic acini. EGF (20 nM) alone had no effect on IP_3-production and amylase release. However, it inhibited CCK-stimulated IP_3-production at 15 seconds after addition of the hormone by $77 \pm 12\%$ and CCK (0.1 nM)-stimulated amylase release by $62 \pm 18\%$ (n = 3). The G-protein inhibitor GDP reduced EGF-effects on CCK-stimulated amylase release and IP_3-production. These results suggest that EGF inhibits CCK-stimulated amylase secretion by activating inhibitory phospholipase C coupling G-proteins.

111 Intra- and Postoperative Serum Levels of TPS and CA 12-5 in Pancreatic Surgery

G. Pohlmann, E. Gross, C. Biermann[a], *R. Klapdor*[a]
Surgery I, General Hospital Hamburg Barmbek; [a]1st Med. Clinic, University Hamburg, FRG

In a prospective study we analyzed the effects of traumatization, hemodilution and regeneration after pancreatic surgery on the serum levels of TPS, CA 12-5 and CA 19-9.

Patients and Methods. In 23 patients (15 carcinomas, 2 adenomas, 5 chron. pancreatitis), 13 resections and 10 drainage procedures were performed. Blood samples were taken 4 h on day 1 (in 4 patients with non resectable carcinomas only preop.) at the end of the operation, every 12 h on days 2 and 3 and on days 5, 8, 13, 21 and 28 postop.

Results. Preop. data: TPS showed the highest sensitivity with 86% for pancreatic carcinomas, CA 19-9 73% and CA 12-5 46.6% (cutoff 55 U/ml, 37 U/ml, 35 U/ml). Combination of TPS and CA 19-9 resulted in a sensitivity for carcinomas of 100%. Intra- and early postop. period: CA 12-5 and CA 19-9 showed a significant decrease of serum levels with a peak decrease after 4–8 h (CA 19-9 x: 64% compared to the preop. values). There was no dependence on disease or preop. serum level, indicating that the decrease is mainly a hemodilution effect. In contrast TPS increased significantly with max. values at end of the operation (x: increase 363% compared to the preop. values). This correlates with the extent of traumatization so that a decrease into normal or preop. values is seen 3–5 days after pancreatic surgery. Postop. period: CA 19-9 reflected with decrease/increase complete tumor resection or tumor progress, CA 12-5 showed another behaviour: a new transient increase after 4–8 days with max. levels after 28 days (x: increase 64%). This effect is seen in all patients, also with normal preop. values. Probably regenerative processes with cell stimulation are responsible for this phenomenon after pancreatic surgery.

Conclusion. The simultaneous preop. measuring of TPS and CA 19-9 results in a sensitivity of 100%. TPS seems to be a new parameter for traumatization of pancreatic tissue. CA 12-5 is probably an indicator for regeneration after pancreatic surgery.

112 The Value of Intraoperative Sonography in Surgery for Chronic Pancreatitis

H. Printz, H.J. Klotter, M. Neurath, M. Rothmund
Department of Surgery, Philipps University of Marburg, FRG

The use of intraoperative ultrasonography (IOUS) is gaining popularity, particularly as a technical aid in pancreatic surgery. Intraoperative localization of the dilated pancreatic duct within the hard, fibrotic and chronically inflamed pancreas merely by bimanual palpation and needle puncture is a difficult and time-consuming task.

We report our experience with intraoperative ultrasonography in 40 patients undergoing surgery for chronic pancreatitis. Among drainage procedures, there were 14 laterolateral pancreaticojejunostomies, 15 pseudocystojejunostomies and 2 pseudocystoduodenostomies. Under the guidance of intraoperative ultrasonography left-sided partial resection of the pancreas was performed in 7 patients, while a Whipple-type procedure was necessary in 6 cases. All preoperatively diagnosed pseudocysts, abscess formations and dilated pancreatic ductal systems could be easily localized with the assistance of intraoperative ultrasound.

Additionally to diagnoses already made preoperatively, intraoperative ultrasonography revealed a second, smaller pseudocyst in one patient and pancreaticolithiasis in another case. However, significant assistance and comfort to the operating surgeon was provided in all cases by intraoperative ultrasound imaging. IOUS assisted to assure the surgeon that associated abnormalities, which were undiagnosed either on preoperative studies or on exploration, were not being missed. This technique, which is cost-effective and minimally invasive, proved to be extremely helpful in localizing pancreatic fluid collections and the course of the pancreatic duct. It facilitates the operation by reducing tissue traumatization and operative time. In conclusion, IOUS is a rapid, safe and useful adjunct to surgical inspection and palpation for diagnosing and defining pancreatic pseudocysts, abscesses and the pancreatic duct.

113 Soluble Interleukin-2 Receptor Serum Levels in Pancreatic Adenocarcinoma (KP) and Chronic Pancreatitis (CP)

P.G. Rabitti, G. Uomo, O.A. Spada[a], *M. Laccetti, A. Del Mastro, B. Marcopido, M. Visconti*
Pancreatic Disease Center and [a]T.E.R.E., Cardarelli Hospital, Napoli, Italy

The increase of soluble Interleukin-2 receptor (sIL-2R) serum levels has been observed in many lymphoproliferative diseases and in some solid neoplasms. Few data are available about sIl-2R serum levels in patients with chronic pancreatic diseases as KP and CP.

We studied the behaviour of sIL-2R in 32 patients with KP (20 men, 12 women; mean age: 57.3 ± 11.3 years) and in 17 patients with CP (17 men; mean age: 53.4 ± 6.2 years). Sera of CP patients were collected out of a pain relapse. All the samples were stored at –20 °C until determination. The sIL-2R levels were detected by Eurogenetic Elisa Kit and expressed in U/ml (reference value <100 U/ml). Each specimen was measured twice and the results were compared using Student's t test.

Results. The mean value in our healthy controls (14 cases) was 119.8 U/ml (± 39.2). The range of values in KP group was 112–809.9 (mean ± SD = 277.8 ± 129.4); the range in CP group was 3.9–226.6 (mean ± SD = 128.5 ± 65.3). Statistical analysis showed a highly significant difference ($p<0.0001$) in the comparison of sIL-2R values of KP versus CP and control subjects. No difference was seen between CP and controls.

Conclusion. These preliminary results point out that the sIL-2R determination could be useful in the differential diagnosis between KP and CP.

114 Conservative Therapy of Necrotizing Pancreatitis

B. Rau, W. Uhl, M. Büchler, P. Malfertheiner, R. Isenmann, H.G. Beger
Department of Surgery, University of Ulm, FRG

There seems to be no doubt as to the role of surgery in extended and infected pancreatic necrosis. The clinical management of patients with minor pancreatic necrosis is still a matter of controversy.

Patients and Methods. Between 5/82 and 5/92, 226 patients with necrotizing pancreatitis (NP) were analyzed. This represents 20% of all the patients hospitalized for acute pancreatitis during this period. 56 patients (39 male, 17 female; mean age 49 years [23–82]; etiology: 61% alcohol, 18% biliary, 21% others) from the group with NP were treated conservatively. The extent of necrosis was diagnosed by Angio CT scan.

Results. The mean Ranson and APACHE II scores in these patients were 2.6 points (range 0–5) and 7.1 (range 0–17), respectively. Angio CT revealed minor (<30%) in 36 and major (≥30–50%) pancreatic necrosis in 11, 6 patients showed extended necrosis (>50%) of pancreatic parenchyma. 52/56 patients responded within 72 h to intensive-care therapy. The most frequent complication was pulmonary insufficiency in 43% (24/56). The mean duration of hospitalisation was 27.5 (7–120) days with a mean duration of ICU treatment of 11.4 days (1–120). The mortality rates were 7.1 and 13.3% in conservative and surgical management of NP, respectively.

Conclusion. Obviously patients with minor and some with major pancreatic necroses recover from severe acute pancreatitis without surgical intervention in response to ICU therapy. However, in these patients pancreatic infection should be carefully excluded by fine needle aspiration.

115 Cholinergic Mechanisms Affect the Exocrine Pancreatic Response and Secretin Release after Intraduodenal Acetic Acid

F. Rodríguez-Nodal, C. Sánchez-Vicente, A. Minguela, L.J. García[a], *J.I. San Román, J.J. Calvo, M.A. López*
Departamento de Fisiología y Farmacología, Universidad de Salamanca; [a]Departamento de Fisiología, Universidad de Extremadura, Spain

Exocrine pancreatic response to the intraduodenal perfusion of short chain fatty acids has received little attention in the past few years, although they can appear into the digestive tract in relatively high concentrations. The aim of this work was to study the effect of intraduodenal perfusion of acetic acid (0.25 *M* at a flow rate of 2 ml/min for 10 min) on exocrine pancreatic secretion of normal or previously atropinized rats. The pancreatic flow rate was determined by weighing, assuming the density of the juice to be 1.0 g/ml. Protein concentration in pancreatic juice was measured by the method of Bradford. Plasma secretin and VIP levels were measured by RIA methods in the portal plasma.

Acetic acid stimulated the hydroelectrolytic fraction of both normal rats (from a mean basal value of 0.29 ± 0.03 μl/min to a maximum of 1.06 ± 0.08 μl/min) and atropinized ones (from a mean basal value of 0.24 ± 0.02 μl/min to a maximum of 0.64 ± 0.12 μl/min). Concerning the protein output a significant increase was also observed in non-atropinized rats from 11.16 ± 2.23 μg/min to a maximum of 35 ± 7.4 μg/min while in the atropinized group a less prominent increase was seen: from 8.78 ± 0.95 μg/min to a maximum of 12.06 ± 2.44 μg/min. This demonstrates the existence of a cholinergic component in the exocrine pancreatic secretion response to intraduodenal acetic acid. Acetic acid did not modify the VIP levels in both normal and atropinized rats, whereas a slight and lasting increase of secretin levels was observed in the non-atropinized group and no increase of secretin release was seen in the atropinized animals. The ANOVA test showed the existence of significant differences ($p < 0.01$) in the plasma secretin release of atropinized and non-atropinized animals.

Although atropine did not influence the secretin release after intraduodenal HCl perfusion it seems to exist a cholinergic component in the release of this peptide depending on the nature of the intraduodenal stimulant as we have just demonstrated with acetic acid and according to that previously reported in the rabbit after intraduodenal alkalinization [García LJ, et al, Exp Physiol 1992; 77:601–613].

116 VLA Receptor Expression and Function in the Human Pancreas

A. Rosendahl, R. Weinel, K. Neumann[a], *D. Erb, M. Rothmund*
Departments of Surgery and [a]Pathology, University of Marburg, FRG

VLA receptors mediate cell adhesion to extracellular matrix and other cells. These interactions are thought to be involved in metastasis and invasive tumor growth. Only limited data are available on VLA receptors in pancreatic carcinoma. Using immunochemistry and adhesion assays we investigated the expression and function of VLA receptors in the normal and malignant human pancreas and pancreatic tumor cell lines.

In the normal pancreas, VLA 2 and VLA 6 were strongly expressed and primarily localized at the basement membrane side of epithelial cells, while VLA 5 and VLA 3 are predominantly expressed on centroacinar cells. Pancreatic carcinoma showed a diffuse distribution of all VLA receptors. The strength of VLA 2 and VLA 6 expression was unchanged, while VLA 5 showed a loss in 50–85% of pancreatic carcinoma cells. The VLA 3 expression in carcinoma cells was heterogeneous, ranging from negative to moderate. In adhesion assays VLA 6 facilitated tumor cell adhesion to laminin, while VLA 2 acted as a receptor for collagen and VLA 5 bound to fibronectin.

While the laminin receptor VLA 6 and the collagen receptor VLA 2 may be responsible for regular cell-basement membrane interaction, the diffuse distribution of VLA 6 and VLA 2 in carcinoma cells might be associated with their ability to randomly interact with extracellular matrix structures during invasion and metastasis. A loss of fibronectin receptors, as we saw with VLA 5 in pancreatic carcinoma, has been reported to be associated with a malignant phenotype of cells.

117 Experimental Evidence Against Opie's Common Channel Bile Reflux Theory

N.S. Runkel, F.G. Moody[a], *W. Müller, N. Senninger, Ch. Herfarth*
Departments of Surgery, University of Heidelberg, FRG, and [a]University of Texas Medical School, Houston, TX, USA

Opie's theory that common bilio-pancreatic channel (CC) obstruction results in acute pancreatitis secondary to bile reflux was tested in 12 opossums. Pressures ($cmH_2O \pm SD$) and fluid movements were assessed in the pancreatic duct (PD) and common bile duct (CBD) before and after CC obstruction using a 'propulsion evaluation system'. Electrodes recorded the sphincter of Oddi (SO) electromyography.

Patent common channel. Baseline saline perfusion of the CBD and the PD (0.03 ml/min each) resulted in a pressure of 18.7 ± 8.2 in the CBD and 18.4 ± 11.9 in the PD. Stepwise increase in perfusion of either CBD or PD to 0.3 ml/min induced a rise of CBD pressure to 43.7 ± 8.1 and PD pressure to 42.8 ± 11.6. The pressure of PD remained at 20.1 ± 8.7, and CBD pressure increased to 29.8 ± 14.0 when alternate ducts were perfused.

Common channel obstruction. Pressures of 50 in PD and <20 in the CBD resulted in a fluid movement from PD to CBD in all animals until pressures equalled. Fluids failed to move when pressures were set at <20 in the PD and 50 in CBD in 10 animals. Fluid flowed in only 2 animals until pressures were equalized. There was no alteration of SO electromyography.

Conclusions. Movements of fluid in the CC are under regulation of the SO whether the CC is patent or not. Acute obstruction causes free reflux of fluid from the PD into the CBD but not from the CBD into the PD in this system. Thus, CC obstruction does not result in bile reflux; Opie's theory is questioned in this experimental model.

Supported by NIH GM38529 and Deutsche Forschungsgemeinschaft Ru 387/1–2*.

118 Lipid Metabolism in Acute Biliary Pancreatitis

M. Sachs, C. Gurlitt[a], *H. Förster*[a], *A. Encke*
Departments of General Surgery and
[a]Anesthesiology, University of Frankfurt/M., FRG

Aim. It has been suggested that the enzymes of lipid metabolism (lipase, phospholipases and sterol ester hydrolase) demonstrable in the blood during acute pancreatitis are pathogenetically relevant since they can release cytotoxic lysolecithins and free fatty acids (FFA). We wanted to investigate whether the pancreatic enzymes released into the blood are able to metabolize lipids.

Methods. Interstitial-edematous pancreatitis was produced in 40 rats (Wistar) by ligature of the common duct close to the duodenum causing an obstruction of the flow of bile and pancreatic juice. Serum was examined after 0 (control), 3, 6, 12 and 24 h using the usual enzymatic and chemical methods: lipase (U/l), cholesterol esters (mg/dl), triglycerides (mg/dl), and FFA ([μmol/l], by gas chromatographic analyses). The pancreas was examined by light microscopy.

Results. Despite high in vitro activity of serum pancreatic enzymes, the serum concentrations of triglycerides, FFA and cholesterol esters remained unchanged during the period of investigation.

$\bar{x}$, (s)		0 h	3 h	12 h	24 h
Lipase	($\bar{x}$)	15	1,559	1,294	466
	(s)	8	627	354	349
FFA	($\bar{x}$)	775	680	613	693
	(s)	63	132	82	64
Triglycerides	($\bar{x}$)	52	79	49	98
	(s)	5	20	12	18
Cholesterol esters	($\bar{x}$)	62	48	56	52
	(s)	8	10	7	16

$p<0.01$.

Conclusions. Although enzyme activity analysis shows the enzymes released during pancreatitis to be biochemically active in vitro, these enzymes obviously have no effects on serum lipids in vivo. Therefore, it seems to be unlikely that enzymes of lipid metabolism play any decisive role in the pathogenesis of acute pancreatitis.

119 Geographical Differences of Ki-ras Mutations in European Pancreatic Carcinomas

A. Scarpa, P. Capelli, G. Zamboni, C. Bassi, S. Hirohashi[a]
Istituto di Anatomia Patologica and Clinica Chirurgica, Verona, Italy; [a]Pathology Division, Natl. Cancer Center, Tokyo, Japan

Nineteen Italian pancreatic adenocarcinomas were tested for the presence and pattern of mutations in codons 12 and 13 and 61 of Ki-ras gene. A powerful method for the detection of structural alterations of DNA was used, namely single-strand conformation polymorphism (SSCP) of PCR-amplified DNA fragments. Frozen tissues, obtained at surgery, were used as source of DNA. Mutations were detected in 16/19 samples (84%) and always involved the second base of codon 12. They were G to A transitions (GGT→GAT; *gly*→asp) in 8 cases and G to T transversions (GGT→GTT; *gly*→val) in the remaining 8 cases.

Our results are in keeping with different studies demonstrating a high frequency of Ki-ras codon 12 mutations in pancreatic carcinomas [1]. Remarkable differences, however, can be noticed when comparing the type of mutations in cancers of different European countries. In fact, in Italian cases two types of mutations were observed, *asp* and *val*. In the two other European studies [2, 3] a third type of mutation was present, *arg* in Austrian and *cys* in Dutch cases. The two mutations found in Italian cancers accounted each for 50% of the cases, whereas in Austrian and Dutch cancers each of the three mutations were present in one-third of the cases.

The observation of geographic differences in the Ki-ras mutational pattern suggests that different environmental and genetic factors are involved in the pathogenesis of pancreatic cancer. The knowledge of the spectrum of gene mutations in cancers of a specific geographic area could help in addressing epidemiological studies seeking for etiologic factors and might reveal helpful in devising new therapeutic approaches.

1 Shibata C, et al: Ballière's Clin Gastroenterol 1990; 4:151.
2 Smit V, et al: Nucleic Acids Res 1988;16:7773.
3 Grünewald K, et al: Int J Cancer 1989;43:1037.

Supported by Associazione Italiana per la Ricerca sul Cancro, Milano, Italy.

120 p53 Gene Mutations in Pancreatic Cancer

A. Scarpa, K. Mukai, G. Zamboni, P. Capelli, S. Hirohashi
Pathology Division, National Cancer Center Research Institute, Tokyo, Japan; Istituto di Anatomia Patologica, Verona, Italy

Ki-ras gene mutations are a feature of pancreatic adenocarcinoma. Very little is known about additional genetic abnormalities in this malignancy. We studied 34 pancreatic adenocarcinomas (20 Italian and 14 Japanese) for the presence of p53 gene mutations by single-strand conformation polymorphism method and by direct sequencing of PCR-amplified fragments.

Mutations were detected in 14 cases. The transitions were 6 G to A and 2 A to G; the transversions were 1 C to G and 2 A to C; the remaining 3 were frameshift mutations. Thus, mutational pattern suggests that p53 gene alterations in pancreatic cancer are most likely due to spontaneous errors during DNA replication rather than to specific carcinogens. No statistically significant association between p53 alterations and clinico-pathologic characteristics of cancers was uncovered. It may be worth reporting, however, the observations that cases with no p53 alterations had a slightly longer median survival with respect to those showing a significant proportion of cells with p53 alterations (13 vs. 8 months).

Also in the light of the demonstrated cooperation of ras and p53 gene alterations in the transformation of cultured cells, our data suggest that p53 mutation is one of the genetic defects, occurring at variable stages of tumor development and progression, that may play a role in the pathogenesis of a proportion of pancreatic cancers.

Supported by the Ministry of Health and Welfare of Japan; and by Associazione Italiana per la Ricerca sul Cancro, Milano, Italy.

121 Effect of S_p- and R_p-Diastereomers of Adenosine Cyclic 3′,5′-Phosphorothioate on Rat Pancreatic Acini

C. Schäfer, H. Printz, B. Göke
Departments of Internal Medicine and Surgery, University of Marburg, FRG

Previous studies established selective inhibitory (R_p-cAMP) or stimulatory (S_p-cAMP) effects of diastereomers of adenosine 3′,5′-phosphorothioate on in vitro activation of cAMP-dependent protein kinase (PKA). We have evaluated their in vivo potency in isolated rat pancreatic acini on the search for highly selective tools to monitor cAMP-dependent PKA activation and its role in amylase exocytosis. In intact acini R_p-cAMP had no effect on the basal or secretin-induced amylase secretion, whereas S_p-cAMP at 1 m*M* slightly enhanced amylase release above basal (1.1±0.1% of total amylase; 20 min incubation). When cells were permeabilized by 10 μ*M* digitonin the S_p-isomer dose-dependently evoked a strong amylase release. This effect was significant at 100 μ*M* and maximal at 1 m*M* (7% of total amylase). R_p-cAMP (10, 100, 1000 μ*M*) dose-dependently inhibited but not completely abolished amylase secretion induced by 0.5 μ*M* secretin (54% inhibition) or 100 μ*M* S_p-cAMP (57% inhibition). Stimulation of secretion by an optimal secretin load (0.5 μ*M*) was further increased by S_p-cAMP (100 μ*M*: 11%, and 1 m*M*: 45% above secretin alone). PKA activity was stimulated 6.6-fold above basal by 100 μ*M* S_p-cAMP, 7-fold by 10 μ*M* forskolin, and only 3.1-fold by 10 μ*M* secretin. In further experiments 8-bromo-S_p-cAMP, a recently developed S_p-cAMP derivative, was tested on non-permeabilized cells. This compound shows an about 100-fold increased lipophilicity and induced a dose-dependent stimulation of amylase secretion (basal: 1.8% of total amylase; 10 μ*M*: 1.7%, 100 μ*M*: 4.9%, 1 m*M*: 5.7%; secretin 10 μ*M*: 12.5%; forskolin 100 μ*M*: 4.5%).

Conclusion. These results prove that cAMP-dependent PKA is involved in the exocytoses of amylase. R_p- and S_p-cAMP are useful tools to selectively modulate PKA activity in permeabilized cells, but only the 8-bromo-substituted derivative permeates into cells in amounts sufficient to alter PKA in intact cells.

(Supported by Kempkes grant 19/91).

122 Glutathione and Energy Metabolism in Acute Experimental Pancreatitis

M.H. Schoenberg, M. Büchler, S. Eisele, M. Younes[a], H.G. Beger
Department of General Surgery, University of Ulm, FRG; [a]Max-von-Pettenkofer-Institut, BGA, Berlin, FRG

The enzyme glutathione peroxidase (GPO) and its cosubstrate glutathione (GSH) are the most important detoxification systems against oxygen radicals (OR). Thereby GSH is converted to its oxidized form (GSSG). As shown previously, OR play an important role in acute pancreatitis (AP). Aim of our experimental study was to assess the changes of GSH and purine metabolism, essential for GSH resynthesis, in the early phase of AP.

Methods. In rats AP was induced by continuous i.v. infusion of cerulein (dosage 0.5 μg/kg/h), NP by retrograde injection of 5% sodium-taurocholate into the pancreatic duct. After 0.5, 3.5 and 12 h the pancreas was removed for determination of ATP, ADP, AMP, GSH and GSSG and for light microscopy. Rats were also treated with the radical scavengers (FRS) superoxide dismutase (SOD: dosage 150,000 U/kg/h) and catalase (CAT: dosage 400,000 U/kg).

Results. Already after 0.5 h the ATP concentration in AP decreased (6.63 ± 0.7 to 2.8 ± 0.3 nmol/mg protein; $p < 0.005$) and remained low throughout the experiment. ADP and AMP did not change. GSH also decreased (3.5 h: 19.9 ± 3.2 to 7.4 ± 0.9 nmol/mg protein; $p < 0.002$) and remained at this level. In contrast the percentage of the oxidized form GSSG as compared to total glutathione increased from 4.1 ± 1.0 to 5.9 ± 1.2%. In NP changes were much more dramatic: ATP decreased after 0.5 h from 6.3 ± 0.7 to 0.6 ± 0.1 μmol/mg protein ($p < 0.0001$) and did also not increase during the experiment. GSH dropped (19.5 ± 2.6 to 9.3 ± 3.6 after 0.5 h), whereas GSSG increased (from 2.7 ± 0.4 to 10.1 ± 1.2%; $p < 0.001$) and remained at this level. In both models these alterations (GSH, GSSG, purine metabolites) preceded the histologic changes observed.

Conclusion. In the early phase of AP and NP, oxidative stress leads to a depletion of GSH at the expense of GSSG. In the later phase of AP, GSH resynthesis is impeded by decreased ATP, thus impairing the detoxification of OR.

123 Late Histopathologic Changes and Healing in an Improved Rodent Model of Acute Necrotizing Pancreatitis

J. Schmidt, D.W. Rattner, K. Lewandroski, C.C. Compton, A.L. Warshaw
Massachusetts General Hospital, Harvard Medical School, Boston, MA, USA

Tissue injury causes local and systemic complications, key determinants of acute pancreatitis. We recently described an improved pancreatitis model in which hyperstimulation with intravenous (iv) caerulein (CAE) is superimposed on very low dose (10 mmol/l) intraductal (id) glycodeoxycholic acid (GDOC) [Ann Surg 1992;215:44–56], which produces a homogeneous distribution of inflammation and necrosis of moderate severity resembling human pancreatitis. The purpose of this study is to evaluate the late (14 days) histopathologic consequences of the acute pancreatic injury.

Methods. 48 Sprague-Dawley rats were assigned to 4 groups ($n = 12$): saline iv/id, CAE iv/saline id, CAE iv/GDOC 10 mmol/l id, and saline iv/GDOC 34 mmol/l id. All id infusions (0.10 ml) were at 30 mmHg over 1.5 min. CAE was given at 5 μg/kg/h for 6 h. Tissue for histopathology obtained at death or sacrifice at 14 days was evaluated independently by two pathologists.

Results. All saline and CAE treated controls survived without complications, and there was no histopathologic difference between these groups. Animals treated with high-dose GDOC-34 (traditional model) had severe acute and chronic inflammation, acinar necrosis, hemosiderin deposits, and a high degree of mitotic activity with scleroductal, acinar and stromal proliferation, all disportionately confined to the pancreatic head. 14-day mortality was 25%. In contrast, low-dose GDOC-10 combined with CAE caused evenly distributed chronic inflammation, scleroductal proliferation, acinar atrophy, and stromal reaction. There were retroperitoneal abscesses (2), multiple intestinal adhesions (10), and renal (6) and pulmonary (6) injury. 14-day mortality was 15.7%.

Conclusion. The combination of low-dose GDOC and CAE produces late histopathologic changes and complications mimicking those of necrotizing pancreatitis in man. This model is well suited to study healing after acute pancreatitis.

124 The Effect of Epinephrine, Norepinephrine and Phenylephrine on the Microcirculation of the Intact Pancreas of the Rabbit

W. Schratt, E. Klar, H. Buhr, Ch. Herfarth
Department of Surgery, University of Heidelberg, FRG

Introduction. Disturbances of the microcirculation of the pancreas are of important clinical relevance especially during acute pancreatitis. The aim of this study was to examine the effect of catecholamines, widely used in intensive care medicine, on the microcirculation of the pancreas.

Methods. By means of intravital microscopy we assessed the microcirculation of the pancreas of the New Zealand white rabbit. The pancreas had been exposed for intravital microscopy through a right lateral laparotomy and fixed on a Plexiglas stage. Using the plasma marker FITC-dextran 70 we analyzed the capillary density as well as blood cell velocity. We analyzed microcirculation under control conditions ($n = 5$) and after i.v. application of epinephrine (3 μg/kg/min) ($n = 5$), norepinephrine (3 μg/kg/min) ($n = 5$) and phenylephrine (10 μg/kg/min) ($n = 5$) up to 3 h after application.

Results. Under control conditions the microcirculation of the exposed pancreas was stable during the 3-h observation period. Despite a significant increase in MAP, phenylephrine and norepinephrine caused a marked reduction of the number of perfused capillaries from 97 ± 4 to 57 ± 17% ($p < 0.005$) after 2 h. Average blood cell velocity in the remaining perfused capillaries also declined from 0.83 ± 0.19 to 0.53 ± 0.15 mm/sec. The effects of epinephrine on the microcirculation were less pronounced and were not of statistical significance, although the increase in MAP was comparable.

Conclusions. α-adrenergic agents impair the microcirculation of the intact pancreas despite the increase in MAP. In further studies the effect of these agents on the microcirculation of the organ during acute pancreatitis has to be examined, as in acute pancreatitis the preservation of the microcirculation seems to be of crucial relevance.

125 Effects of Free Radical Scavengers on Oxidative Stress-Induced Injury to Pancreatic Acinar Cells

H.-U. Schulz, C. Niederau, G. Letko
Department of Surgery, Medical Academy of Magdeburg, and Department of Gastroenterology, Heinrich Heine University, Düsseldorf, FRG

Peroxidation of membrane lipids contributes to cell injury in acute and chronic pancreatitis. The present study was designed to investigate (1) the susceptibility of exocrine pancreatic cells to oxidative injury and (2) the potential of free radical-scavenging enzymes to prevent such damage.

Methods. Freshly isolated rat pancreatic acinar cells (10^6 cells/ml) were exposed to two different models of oxidative stress. Membrane-damaging oxygen-free radicals were generated by a reaction mixture containing xanthine oxidase (XO; 20 mU/ml), hypoxanthine (HX; 500 μmol/l), and EDTA-chelated $FeCl_3$ (50 μmol/l) dissolved in Krebs-Henseleit buffer (KHB). In a second model, acinar cell injury was initiated by lipid peroxidation using adenosine-5′-diphosphate (ADP; 5.3 mmol/l) and $FeSO_4$ (300 μmol/l) dissolved in KHB. Cell damage was assessed by both cellular trypan blue (TB) exclusion and release of lactate dehydrogenase (LDH). At the time point at which cell damage in the ADP/Fe^{2+}- and XO/HX-groups containing no scavengers reached 50%, cell injury was determined in all other groups.

Results. TB exclusion indicated cell damage earlier and more sensitive than LDH release. XO/HX-induced acinar cell injury was effectively prevented by catalase (CAT) and glutathione peroxidase (GPx), whereas superoxide dismutase (SOD) enhanced cell injury. SOD, however, was the only enzyme capable of reducing ADP/Fe^{2+}-induced cell damage:

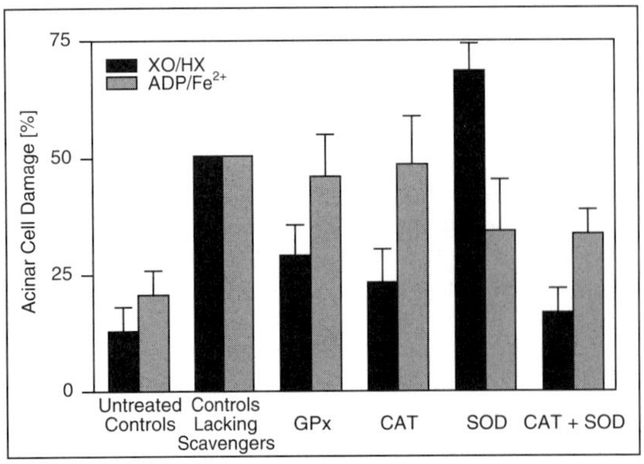

Conclusions. Pancreatic acinar cells are very susceptible to damage by oxidative stress. Different radical species and multiple mechanisms seem to be involved in this process. Some classical scavengers such as SOD may rather damage than protect acinar cells when H_2O_2 is generated in concentrations overwhelming the capacity of endogenous catalase. Therapeutic approaches to pancreatic disease using antioxidants should therefore include combinations of protective substances. The present in-vitro model may help to further clarify the mechanisms of radical-induced damage to the pancreas and to plan meaningful in-vivo studies.

126 Immunophenotyping of Intraductal Pancreatic Tumors: Evidence of Gastrointestinal Differentiation

F. Sessa, O. Luinetti, E. Solcia, G. Klöppel
Departments of Pathology, IRCCS, S. Matteo Hospital and University of Pavia, Italy, Academic Hospital Jette, Free University of Brussels, Belgium

The group of pancreatic tumors with intraductal growth encompasses neoplasms that have been described as intraductal papillary (IPT) and intraductal mucin-hypersecreting tumor (IMHT). The two are histologically indistinguishable and differ only for the viscous mucin stuffing dilatated pancreatic ducts of IMHT. The aim of our study was to characterize the mucinous content and the epithelial cell growth. Nine tumors (5 IPT and 4 IMHT) were studied with antibodies against markers of gastric glands (pepsinogen II), of gastric superficial-foveolar cells (M1 fucomucins and catepsin E), intestinal colorectal epithelium (CAR-5 mucin like antigen), intestinal goblet cells (M3 mucin antigen) and pancreatic duct epithelia (mucin like DUPAN-2 antigen).

Results. Pg II was found in 3 IPT and 1 IMHT, CaE in all tumors, M1 in 3 IPT and in 3 IMHT, CAR-5 in 4 IPT and in 3 IMHT, M3 in 2 IPT and in 4 IMHT. Dupan-2 was found in a small percentage of cells in 1 IPT and 1 IMHT, respectively. The two intestinal markers were coexpressed in 3 IMHT and only in 1 IPT. Gastric markers of mucous gland and superficial cell were present in 4 IPT. Our results suggest that the two tumors are not related to normal duct epithelial cells in lacking DUPAN-2 immunoreactivity. Their large expression of markers of gastrointestinal cells suggest a possible relationship with pancreatic duct metaplastic lesions that share the same markers and are lacking pancreatic markers. Finally, intestinal differentiation seems to be more prominent in IMHT than in IPT where gastric differentiation is more prominent. The two tumor types could represent an entity expressing all the spectrum of gastro-intestinal epithelial cells.

127 Changes of High Energy Phosphates in a Model of Experimental Alcoholic Pancreatitis versus Taurocholate-Induced Pancreatitis in Rats

M. Siech, Ch. Sotak
Department of General Surgery, University of Ulm, FRG; Worcester Polytechnic Institute, Worcester, MA, USA

Acute pancreatitis (AP) is known to be associated with changes in the energy supply of the pancreas, and hypoxia might be a contributing factor in this disease as well. Consequently, we have been investigating the use of in vivo ^{31}P NMR spectroscopy in a new alcoholic pancreatitis model to evaluate pancreatic hypoxia after the induction of AP in the rat compared to the model of taurocholate (TC)-induced AP. Therefore the pancreas was mobilized by a technique reported earlier allowing to place corpus and tail of the organ in a radiofrequency coil tuned to 34.63 MHz. ^{31}P spectra were continuously obtained from the pancreas on GE 2.0 Tesla imaging spectrometer.

Alcoholic acute pancreatitis was induced by ethanol intoxication and a reversible obstruction/hypersecretion mechanism as recently reported. TC-pancreatitis was induced by instillation of 5 or 10% TC, respectively. In latter cases ATP, was decreasing, anorganic phosphate (Pi) was irreversibly increasing, and histologically severe inflammation and extensive hemorrhage of the organ were found. However, in 5% TC there were no hemorrhage and no change of the high energy phosphates. In alcoholic AP, which is a mild form of inflammation with fat and parenchymal necrosis but without hemorrhage, there were neither decreases in ATP nor increase of anorganic phosphate. The results imply that a very poor energy situation of the pancreas is always due to the hemorrhage component of the disease.

128 Characterization of New Monoclonal Antibodies Directed Against Normal Human Exocrine Pancreas and Pancreatic Adenocarcinomas

H.-P. Sinn, S.A. Brown, E. Oberle, J.S. Thompson
Department of Pathology, University of Heidelberg, FRG, and Department of Medicine, University of Kentucky Medical Center, Lexington, KY, USA

A series of 12 cell fusions of BALB/c mice spleen cells with the SP2/O-Ag14 mouse myeloma cell line were performed after immunizations with normal human pancreas in order to generate monoclonal antibodies (mAb) with high affinity against antigens of the exocrine pancreas. Clones were characterized histologically and with lectin-binding assays. The immunohistologic tissue screening resulted in the selection of fourteen clones that were further characterized on frozen tissues, tumor cell lines and lectin binding assays. Four out of seven mAb reacting with pancreatic acini detected granular antigens and three mAb reacted with cell membrane antigens. Four other clones reacted with pancreatic ducts, the remaining three clones detected antigens present both in acini and ducts. Among the clones reactive with secretory products, there was one pair of mAb identified as pancreas-specific (P78C9/P79D4), the other secretory antigens (identified by P97F3 and P109H1) were also detectable in other exocrine organs, but absent from the gastrointestinal mucosa. Two clones reactive with acinar cell membranes (P96H2, P100H1) were also positive with hepatocytes. All but four antigens were detectable in pancreatic cancer cell lines Capan-1, Capan-2 or DAN-G. On the whole, immunizations were more effective using whole pancreatic tissues (8 clones selected) compared to single cell suspensions as immunogen (4 clones) or cell membrane preparations (2 clones). Therefore, the immunization and screening strategies used in this study resulted in the generation of new monoclonal antibodies against specific substructures of the exocrine pancreas. The shared expression of some of these antigens with other exocrine organs and pancreatic adenocarcinoma cell lines suggests that these mAb detect antigens that are characteristic for the exocrine system and could therefore be useful for the study of exocrine organs and adenocarcinomas that are derived from them.

129 Presence of the X-Hapten in Pancreas and Pancreatic Carcinomas

H.-P. Sinn, S.A. Brown, E. Oberle, J.S. Thompson
Department of Pathology, University of Heidelberg, FRG; Department of Medicine, University of Kentucky Medical Center, Lexington, KY, USA

The X-hapten (Lewis^x) epitope consists of the terminal oligosaccharide Galβ1–4(Fucα1–3)GlcNAc-R which is apparently not regulated by the Le or Se genes of the Lewis system. This study was designed to compare the expression of the Lewis^x epitope in normal pancreas with its expression in pancreatic adenocarcinomas and tumor cell lines and to characterize the nature of the antigens bearing this epitope with immunochemical methods. On frozen tissue sections of normal pancreas, the Lewis^x epitope was detected focally in acinar cell granules; sialosyl-Lewis^x was detected in centroacinar cells and the cytoplasm of intralobular ducts. 16 of 19 of pancreatic carcinomas expressed the Lewis^x antigen on tissue sections; however, there was no correlation with prognostic parameters such as tumor grade or stage. 18 of 19 of tumor specimens were positive for sialosyl-Lewis^x. Analysis of pancreatic carcinoma cell lines (CAPAN-1, CAPAN-2 and DAN-G) revealed intracytoplasmatic expression of sialosyl-Lewis^x epitopes in these cells, and cell surface reactivity was detected on flow cytometry for sialosyl-Lewis^x. Lectin-affinity chromatography of cytoplasmatic preparations showed that the Lewis^x positive antigens of normal human pancreas can be eluted using SBA and UEA-I lectins but have little or no affinity to WGA, GS-I, or DBA lectins, indicating the presence of Fuc and Gal oligosaccharide residues. It is concluded that the Lewis^x and sialosyl-Lewis^x antigens are non-polymorphic carbohydrate determinants which are present in secretory granules of acinar cells, ductules and pancreatic secretions. This makes the Lewis^x antigen suitable for the analysis of the secretory process in the exocrine pancreas and the study of secretory differentiation antigens in ductal pancreatic carcinoma.

130 Expression of a Novel Pancreatic Acinar Mucin Defined by Monoclonal Antibody P109H1 in Normal Tissues and Malignant Tumors

H.-P. Sinn, G. Sidaras, S.A. Brown
Department of Pathology, University of Heidelberg, FRG, and Department of Medicine, University of Kentucky Medical Center, Lexington, KY, USA

Mucins and other high molecular weight glycoproteins are commonly expressed as tumor-associated antigens of pancreatic adenocarcinomas. In order to investigate the tissue distribution of a novel pancreatic mucin, defined by the monoclonal antibody P109H1, we tested a variety of tissues on frozen sections for the expression of the P109H1 antigen using the biotin-avidin-complex technique.

In normal fetal and adult pancreas the antigen was restricted to the apical cytoplasm and the cell surface of pancreatic acini and also present in terminal duct secretions. However, it was never detected in ductal cells of the pancreas. Correspondingly, 20 pancreatic adenocarcinomas of the common ductal type and pancreatic tumor cell lines of ductal origin (Capan-1, Capan-2, DAN-G, AsPC-1) were not reactive, while 1 pancreatic carcinoma of probable acinar cell origin was diffusely positive. In the mammary gland, 5 out of 6 invasive-lobular carcinomas also expressed the antigen, corresponding to terminal gland differentiation. Similarly, 7 out of 11 malignant pleural effusions from mammary carcinoma patients were positive in an indirect immunoassay. The tissue specificity of the P109H1 antigen is underlined by the fact that other tissues, such as stomach, small and large intestine, gallbladder or liver, were consistently negative for the presence of the antigen. Also it was not detectable in a series of 25 adenocarcinomas from colon or in stomach cancers.

In conjunction with the biochemical data, it is concluded that P109H1 detects a novel tumor-associated mucin-like glycoprotein of pancreatic acini.

131 Characterization of Carbohydrate Groups on Zymogen Granule Membrane Proteins and Investigation of the Glycosyl Residues Near the Active Sites of the Nucleoside Phosphatases

M.C. Spaans, R.W. Ammann, A.U. Freiburghaus
Department of Internal Medicine, University Hospital, Zurich, Switzerland

Secretion of digestive enzymes involves zymogen granules in the pancreatic acinar cells. This activity may arise from nucleoside phosphatases as part of energy-utilizing processes and the specific attachment of macromolecules to carbohydrate-bearing molecules, which are present at the outer surface of the zymogen granule membrane. To characterize the carbohydrate groups on the few integral proteins present in the membranes of pancreatic zymogen granules (ZMG), we used a labeled avidin-biotin system with biotinylated lectins and avidin conjugated to horseradish peroxidase. Proteins of the ZGM were subjected to non-denaturing electrophoresis in a 5–15% gradient polyacrylamide gel containing 0.2% CHAPS and 0.1 *M* TRIS/HCl, pH 9.5. At least 4 distinct protein bands were detected, all derived from glycoproteins as measured with the PAS assay. The proteins were transferred from electrophoresis gels to nitrocellulose membranes and incubated with biotinylated lectins (Con A, WGA, SBA and RCA 120). Then avidin conjugated to horseradish peroxidase was added which binds to the biotinyalated lectins. 3,3'-Diamino benzidine tetrahydrochloride (DAB) was used as substrate for horseradish peroxidase which in case of lectin binding gives an orange color. In addition, the presence of glycosyl residues near the active sites of nucleoside phosphatases has been investigated by preincubating the gel stripes with different lectins (Con A, WGA, SBA, and RCA 120). After incubation with different substrates (AMP, ADP, and ATP) the nucleoside phosphatase activities were detected by visualization of the resulting lead phosphate precipitates in the gel, which formed in the presence of $Pb(NO_3)_2$. Precipitates became visible after conversion with ammonium sulfide into brown, insoluble PbS. The various lectins bind to various glycoprotein bands and inhibited nucleoside phosphatase activity towards different substrates. In conclusion, the zymogen granule membranes contain at least three glycoprotein bands with nucleoside phosphatase activity and with different substrate and inhibitor specificity. Their function during exocytosis remains to be elucidated.

132 Diabetes and Pancreatic Cancer

C. Sperti, C. Pasquali, R. Behboo, R. Isoardi, F. Cappellazzo, S. Pedrazzoli
Clinica Chirurgica I, University of Padua, Italy

Diabetes frequently occurs in association with or as a consequence of pancreatic cancer. In this study we evaluated the incidence, the prognostic implication and the course of diabetes in pancreatic cancer patients. From 1980–1990, 200 patients with pancreatic cancer to were admitted to our institution.

On the basis of history, blood glucose and glucose tolerance test, the patients were divided as follows: (1) non-diabetic (no-DM = 92; 53%); (2) old-DM (21; 12%); (3) new-DM (61; 35%); (4) not classified (26). Age, sex, jaundice, symptoms and site of the tumor were similar in all groups, although involvement of the whole gland was more frequent in DM groups (10 vs. 1%). The incidence of distant metastases and resection rate were similar among the groups, but the rate of stage I was lower in patients with new-DM than old-DM and no-DM (1.6 vs. 9.7%). Morbidity was not different among the groups, but operative mortality was higher in new-DM and no-DM (8.2 and 12%, respectively) than in old-DM (0%). Long-term survival was significantly better in no-DM and old-DM than in new-DM ($p<0.05$). The follow-up of 33 patients who underwent subtotal pancreatectomy showed that glucose tolerance test improved in 6 cases, was unchanged in 10, while 17 no-DM patients continued to have a normal glucose metabolism after pancreatic resection. In conclusion, new-DM in pancreatic cancer seems to be associated with more advanced stage and less favorable prognosis. Improving of diabetes after pancreatectomy suggests a direct effect of pancreatic cancer on glucose metabolism.

133 Survival and Patterns of Failure after Resection for Pancreatic Carcinoma

C. Sperti, C. Pasquali, F. Cappellazzo, B. Bonadimani, S. Pedrazzoli
Clinica Chirurgica I, University of Padua, Italy

In this study we analyzed survival and sites of recurrence after resection in patients with pancreatic cancer. From 1970 to 1989, 88 patients underwent resectional surgery. There were 38 standard Whipple resections, 21 extended subtotal, 13 total and 16 distal pancreatectomies. Fourteen cases had associated vascular resections and 10 pylorus-preserving pancreatectomies. Overall 5-year survival rate was 14.5%: no statistical difference was found in survival rate after Whipple resection and extended subtotal pancreatectomy ($p=0.08$) or vascular resection ($p=0.15$). Eleven patients are alive and free of disease from 12 months to 15 years. The pattern of recurrence was analyzed in 46 patients who died for progression of disease. Local recurrence was found as the only site in 16 patients (34%) and as component of failure in 63%; liver involvement in 32% and as component in 58%. Five patients developed distant metastases, but only 2 as the single site of failure. No relationship was found between site of failure and type of resection procedure; but local recurrence was the site of relapse in all patients who underwent distal pancreatectomy. Survival after resection for pancreatic carcinoma seems not to be related to the extension of pancreatic resection. Local and hepatic failure rates are very high and strongly suggest the need of strategies to improve loco-regional control after surgical resection.

134 Study on Digestive Function after Pylorus-Preserving Pancreatoduodenectomy (PPPD)

C. Sperti, C. Pasquali, F. Cappellazzo, S. Pedrazzoli
Clinica Chirurgica I, University of Padua, Italy

PPPD has been introduced in 1978 by Traverso and Longmire to prevent gastric resection risk and functional problems observed after pancreatic resection using standard procedures. The aim of the study was to evaluate long-term results of PPPD on digestive function. In 11 patients who underwent PPPD, 3–36 months after surgery, the following investigations were performed: gastroscopy, 24-h pH monitoring, gastric acid secretion test, test meal for gastrin determination, gastric emptying time scan, exocrine pancreatic function tests, OGTT and HIDA scan. Five patients had a pancreatic cancer, 2 ampullary carcinomas and 4 chronic pancreatitis. Pancreatic function tests showed in 9/11 patients with PPPD a marked exocrine insufficiency (in 2 cases it was slight). OGTT performed in 8 non-diabetic PPPD showed an impairment of glucose tolerance in 5 cases. Histology showed a slight gastritis even in 7/11 PPPD patients with normal EGDS; 3 patients had biliary gastric reflux. Gastrin response to test meal was normal and acid secretion was normal in 6/11. Gastric emptying time was normal in 5, slow in 2 and fast in 4 cases. Twenty-four-h gastric pH monitoring showed frequent biliary refluxes in 5 patients as confirmed by HIDA scintiscan. While all PPPD patients were asymptomatic, pylorus preservation does not prevent biliary reflux, and despite a normal EGDS a slight gastritis may occur in up to 50% of cases.

135 DNA Ploidy as a Prognostic Index in Resected Pancreatic Cancer: Preliminary Report

C. Sperti[a], *C. Pasquali*[a], *A. Perasole*[b], *S. Pedrazzoli*[a]
[a]Clinica Chirurgica I, University of Padua; [b]Pathology, O.C. Montebelluna, Italy

DNA content has been demonstrated to provide significant prognostic information in several tumors. Few date about pncreatic cancer have been reported. We analyzed retrospectively 19 specimens of resected pancreatic carcinomas by flow cytometry to assess the rate of aneuploidy and the relationship between survival and DNA patterns. Fifty-micrometer sections of primary tumors and positive lymph nodes were prepared and analyzed by FACS flow cytometer. CV of the G0–G1 peaks ranged from 6.2 to 11.0 (X = 8.0, median 7.9). Eight of 19 (42%) of cases were aneuploid. Out of 9 cases with lymph node metastases, 5 were diploid and 4 aneuploid. Lymph nodes ploidy in 4/9 cases differed from the primary tumor.

Patients with diploid tumors had an overall better prognosis (X = 20 months) vs. aneuploid group (X = 12 months) ($p = 0.001$).

Our small series showed that DNA ploidy may have prognostic value in resected pancreatic carcinoma, however, it is not a reliable index of metastatic lymph node involvement.

136 Immunomodulation of Pancreatic Cancer in vitro. Effect of Monoclonal Antibodies, GM-CSF and Lymphokine-Activated Killer Cells

L. Staib[a, b], *K.H. Link*[a], *M.S. Mitchell*[b]
[a]Department of General Surgery, University of Ulm, FRG; [b]Departments of Medicine and Microbiology, University of Southern California, Los Angeles, CA, USA

Pancreatic cancer is a human malignancy without effective adjuvant chemo- or radiation therapy. Therefore, the search for new therapeutic approaches, such as immunotherapy, is justified. We tested in vitro if (a) pancreatic cancer cells and (b) lymphocytes could be modulated immunologically, in order to yield a cytotoxic effect on pancreatic cancer cells.

Methods. (1) A human pancreatic cancer cell line (PMH, 10^4/ml) was incubated with monoclonal antibodies (BW 494/32 IgG1 and IgG2a) that are directed against pancreatic mucin. Binding of monoclonal antibodies to PMH was tested by fluorescence-activated cell sorting (FACS). Human mononuclear blood cells (effector cells, 10^6/ml) of normal donors were activated by a 2-h incubation with granulocyte-macrophage colony stimulating factor (GM-CSF, 1 μg/ml). Activated effector cells were tested for their cytotoxicity against PMH in a 4-h chromium release assay. (2) Lymphokine-activated killer cells (LAK) were generated by incubation of human normal donor lymphocytes with 1,000 U/ml recombinant human interleukin-2 (IL-2) for 7 days and tested for their cytotoxicity against PMH in a 4-h chromium release assay.

Results. (1) Binding of BW 494/32 IgG1 and IgG2a to PMH was 97 and 98%. Antibody-dependent cytotoxicity of effector cells (without GM-CSF) against PMH was 38 ± 4%, after GM-CSF activation 94 ± 5%. Highest cytotoxicity was obtained with BW 494/32 IgG1. (2) LAK cells generated a dose-dependent cytotoxicity (95, 87, 75 and 34%) at different effector:target ratios (50:1, 25:1, 12:1 and 6:1) against PMH.

Conclusion. Pancreatic tumor cells and effector cells can be immunologically modulated in an effective way with growth factors, cytokines and monoclonal antibodies to inhibit pancreatic cancer cell growth. These in vitro observations may be useful for clinical combination immunotherapy studies.

137 Role of G-Proteins in Bombesin-Stimulated Signal Transduction in Rat Pancreatic Acinar Cells

D. Stryjek-Kaminska, S. Zeuzem, A. Piiper, B. Lembcke, W.F. Caspary
Department of Internal Medicine, Division of Gastroenterology, University of Frankfurt/M., FRG

In pancreatic acinar cells calcium-mobilizing secretagogues stimulate enzyme secretion by activating phospholipase C (PLC). PLC hydrolyzes phosphatidylinositol 4,5-bisphosphate to diacylglycerol and inositol 1,4,5-trisphosphate (IP_3). Heterotrimeric GTP-binding proteins (G-proteins) probably couple hormone receptors to PLC. Bombesin is a growth factor that potently stimulates pancreatic acinar enzyme secretion via activation of PLC. In this study, isolated digitonin-permeabilized rat pancreatic acinar cells were used to investigate the role of G-proteins in bombesin receptor-mediated signal transduction. Weakly hydrolyzable GTP analogs such as guanosine 5′-[3-thio]-trisphosphate (GTP[S]) activate G-proteins, whereas GDP or 2′-deoxy-GDP (dGDP) are inhibitory. GTP[S] increased bombesin (10 n*M*)-induced IP_3 production at 5 sec after addition of the secretagogue by 580 ± 45% (n = 4), while dGDP decreased bombesin (100 n*M*)-induced IP_3production by 111 ± 21% (n = 3). Similarly, GTP[S] increased bombesin (0.1 n*M*)-stimulated amylase release by 154±35% (n=4) and dGDP decreased bombesin (10 nM)-stimulated amylase release by 43±8% (n=3). We therefore conclude that bombesin receptor-mediated signal transduction involves G-proteins in the exocrine pancreas.

138 The Influence of Cecostomy and Colonic Irrigation on the Prognosis of Acute Experimental Pancreatitis

U. Sulkowski, J. Brockmann, P. Dinse, B. Sprakel
Department of General Surgery, University of Münster, FRG

In a controlled trial the influence of cecostomy and colonic irrigation on prognosis and various biochemical parameters was evaluated. Thus, after inducing acute pancreatitis in male Whistar rats by injecting a 2% taurocholate solution into a temporarily closed duodenal loop, 40 animals underwent cecostomy (group B), in 36 the colon was irrigated with normal saline (group C), and 36 served as controls (group A). Altogether, 12 animals succumbed intra- or immediately postoperatively. As a consequence 31 remained in group A, 36 in group B and 33 in group C. Before the end of scheduled follow-up (15 days) the mortality rate in the control group was 22.6%, 16.7% in group B, and 9.1% in group C with the difference between groups A and C being statistically significant at $p<0.05$. This difference was parallelled by a significant difference of serum endotoxin levels between the groups (A 219 ng/l, B 79.2 ng/l, C 71.7 ng/l). Our results suggest that endotoxin might play an important role in the pathophysiology of acute pancreatitis and that endotoxin absorption might be reduced by cecostomy or colonic irrigation.

139 Role of Cystine Transport and Glutathione in the Defence Against Oxidative Stress in Acute Pancreatitis

J.H. Sweiry[a], H.P. Elsässer[b], G.E. Mann[a]
[a]Biomedical Sciences Division, King's College, London, UK; [b]Phillips University, Marburg, FRG

Pancreatic glutathione (GSH) levels are markedly reduced in caerulein-induced pancreatitis [Mann et al., Digestion 1991;49:38; Neuschwander-Tetri et al., J Clin Invest 1992;89:109–116], and treatment with antioxidants and/or GSH prodrugs appears to protect against cellular injury [see Braganza, Curr Opin Gastroenterol 1990;6:763–768; Schoenberg et al., Gut 1990;31:1138–1143; Neuschwander et al., 1992]. In many cells cyst(e)ine transport appears to regulate cellular GSH levels [Bannai and Tateishi, J Membrane Biol 1986;89:1–8]. As defects in ductal function may also underlie the pathology associated with acute pancreatitis [Argent and Gray, Epithelia: Advances in Cell Physiology and Cell Culture, pp 69–67; Kluver, 1990], we have examined the effects of oxidative stress on cystine transport and GSH levels in a human pancreatic ductal cell line (PaTu8902).

PaTu8902 cell monolayers were cultured in DMEM and then incubated in medium containing the oxidant *tert*-butyl hydroperoxide (*t*-bu-OOH, 0.5 mM, 1 h). Transport of L-cystine (10 µM, 5 min) was saturable and inhibited by 0.5 mM L-glutamate. Although *t*-bu-OOH caused no morphological changes, transport of cystine increased from 7.9±0.5 to 12.8±1.4 nmol/mg protein/h (n=6, $p<0.002$). Exposure to *t*-bu-OOH (100 µM) for 24 h decreased GSH levels from 64 to 32 µmol/10^6 cells. Effects of *t*-bu-OOH could not be attributed to an increase in membrane permeability, since uptake of the extracellular marker D-[^{3}H]mannitol remained negligible (1±1 pmol/mg protein/h). Our findings suggest that the pancreatic cystine transporter provides a regulatory mechanism for sustaining cellular antioxidant levels during oxidative stress in acute pancreatis.

(Supported by Ono Pharmaceutical Co., Osaka, Japan.)

140 Is the 24-h Ranson Score a Reliable Prognostic Criterion for Acute Pancreatitis (AP)? A Disappointing Retrospective Review of 111 Consecutive Patients

C. Talamini[a], C. Bassi[b], P. Bovo[a], B. Vaona[a], V. Di Francesco[a], L. Frulloni[a], P. Pederzoli[b], M.P. Brunori[a], M. Filippini[a], C. Andreaus[a], G. Cavallini[a]
[a]Clinica Medica and [b]Clinica Chirurgica, University of Verona, Italy

To qualify as a criterion of prognostic value in AP any clinical, laboratory and/or radiological finding must provide significant indications in terms of mortality and morbidity within 24 h of observation at the latest. To assess whether or not the 24-h Ranson score possesses these requisites, we evaluated it retrospectively in 111 consecutive patients with AP, observed in the first instance in our department, correlating it with (1) type of disease (edematous or necrotic), (2) need for surgical therapy, and (3) mortality. The diagnosis was reached on the basis of routine clinical, laboratory and instrumental criteria. Data were analyzed statistically using the Kendall Tau B, $\chi^2$and logistic regression tests. Fifty-seven patients (51.4%) were suffering from necrotic AP and 54 (48.6%) from edematous AP. Surgical therapy proved necessary in 52 cases (46.8%). The overall mortality was 9.9% (11 patients). The Ranson score was 0 in 35 patients (2 deaths), 1 in 40 (5 deaths), 2 in 20 (3 deaths), 3 in 11 (1 death) and 4 in 5 (no deaths). None of the tests used was capable of establishing any correlation between 24-h Ranson score and severity of disease (edematous or necrotic), type of treatment given (medical or surgical) or mortality. We conclude that the 24-h Ranson score cannot be regarded as a reliable prognostic criterion in AP.

141 Is Cigarette Smoking a Risk Factor in So-Called Alcoholic Pancreatitis (AP)?

G. Talamini, P. Bovo, B. Vaona, L. Rigo, L. Frulloni, V. Di Francesco, M.P. Brunori, A. Castagnini, G. Angelini, P. Pederzoli[a], *G. Cavallini*
Clinica Medica and [a]Clinica Chirurgica, University of Verona, Italy

It is not clearly defined if cigarette smoking plays a role as a risk factor in AP. The aim was to compare drinking and smoking habits in 3 groups of male subjects, all with alcohol intake >40 g/day. 'AAP' group: 67 patients with acute alcoholic pancreatitis, without other known potential causative agents. 'CAP' group: 396 patients with chronic alcoholic pancreatitis. 'C' group: 265 control subjects selected at random from the Verona polling lists and submitted to medical check-up. Variables considered were age at onset of disease, years of drinking and smoking, daily alcohol intake in grams and number of cigarettes smoked daily. Group C differed from the two pancreatitis groups in grams of alcohol/day (C 87.6±33; AAP 154.9±88; CAP 167.4±102) (Mann-Whitney U test $p<0.00001$), while there was no significant difference between AAP and CAP. Similar results were found for the number of cigarettes smoked daily (C 7.9±11; AAP 21.8±17; CAP 22.6±13) (Mann Whitney U test $p<0.00001$). On comparing group C with group AAP+CAP, logistic regression analysis revealed an increased risk of pancreatitis correlated to both higher alcohol intake ($p<0.00001$) and cigarette smoking ($p<0.00001$), while no significant interaction between the two variables was noted, indicating that the two risks are independent.

Conclusion. A higher number of cigarettes smoked daily is a distinct risk factor in AP.

142 Is a Diminished Intestinal Absorption of Tryptophan (TRP) by the M1-Receptor Antagonist Telenzepine (Tel) Responsible for the Inhibited Pancreatic Response to Intraduodenal (ID) TRP?

S. Teyssen, C. Haurand, B. Viell[a], *M.V. Singer*
Department of Medicine, University Hospital of Heidelberg at Mannheim; [a]II. Department of Surgery, University Hospital of Köln, FRG

In 12 conscious dogs with gastric and pancreatic fistulas, dose-response studies of pancreatic protein output to i.d. administration of graded loads of TRP (0.12–10.0 mmol/h) and i.v. 0.15 *M* NaCl (control), given against a background of secretin (S; 20.5 pmol/kg BW/h, i.v.) were done. On another day S plus Tel (243 nmol/kg BW/h, i.v.) was given instead of NaCl. After 1 h of S or S+Tel, i.d. infusion of TRP was begun. TRP plasma levels were determined by HPLC.

Results. Except for 0.12 mmol/h, all subsequent loads of TRP significantly ($p<0.05$) increased the pancreatic protein output over that seen with S alone. Tel significantly inhibited the integrated (225 min) protein output by 87% (S: 22231±6277 mg/225 min; S+Tel: 2947±1543 mg/225 min) to all loads of TRP as compared to experiments without Tel. Tel abolished the response to lower loads (up to 1.1 mmol/h) of TRP; only high loads of TRP (3.3 and 10.0 mmol/h) significantly increased pancreatic protein output over that seen with S+Tel alone. 1.1 mmol/h and higher loads of TRP significantly increased plasma levels of TRP; Tel caused no significant alteration.

Conclusions. These findings indicate that in the intact animal (1) the M1-receptor antagonist Tel is capable of inhibiting pancreatic protein response to i.d. TRP with secretin background (2) at least some cholinergic fibers of the enteropancreatic vagovagal reflex [Am J Physiol 1980;238:G23] have muscarinic receptors of the subtype M1, (3) duodenal absorption of TRP is not altered by Tel, (4) the postabsorption control of pancreatic enzyme secretion by TRP is probably of no or minor importance.

143 Effect of Pancreatic Surgery and Ischemia in a Porcine in Vitro Model on PAF Tissue Levels

A. Tytko, U. Leonhardt, F. Seidensticker, M. Barthel, H. Köhler, W. Creutzfeldt
Department of Medicine and Surgery, University of Göttingen, FRG

Platelet activating factor (1-O-alkyl-2-acetyl-sn-glycero-3-phosphocholine) is an alcyl phosphoglyceride synthesized in exocrine pancreatic tissue, which has recently been shown to be involved in the pathogenesis of experimental pancreatitis. Since acute pancreatitis can be a complication of pancreatic surgery and in particular of pancreas transplantation, the aim of this study was to investigate the PAF immunoreactivity (PAF-IR) in tissue during pancreas ischemia and transplantation. The PAF-IR in tissue was quantified by a commercially obtained radioimmunoassay (Du Pont de Nemours GmbH, Dreieich, FRG). During pancreatectomy the PAF-IR in tissue increased significantly from 142 ± 21 pg/g immediately after laparotomy to 369 ± 53 pg/g ($n = 12$ each, $p < 0.005$) after complete preparation of the pancreas. After pancreas preservation with HTK (Histidine-Tryptophan-Ketoglutarate solution) PAF-IR was reduced to 272 ± 30 pg/g ($n = 12$, $p = 0.125$) but remained significantly higher than after laparotomy ($p < 0.005$). During subsequent cold ischemia in HTK PAF-IR remained between 227 and 308 pg/g tissue ($n = 6$ for each time). These results demonstrate an increase of PAF-IR in tissue during pancreatic surgery, so that the addition of a potent PAF-antagonist to the preservation solution might be beneficial for pancreas transplantation.

144 Interleukin-6 – A Promising Parameter for the Early Assessment of the Severity of Acute Pancreatitis

W. Uhl, M. Büchler, P. Malfertheiner, F.C. Torab, R. Isenmann, H.G. Beger
Department of Surgery, University of Ulm, FRG

In a prospective study we analyzed the value of interleukin-6 and TNF in comparison with CRP for the early assessment of the severity of acute pancreatitis.

Patients and Methods. 15 consecutively hospitalized patients with acute pancreatitis (10 male, 5 female, mean age 51; range 30–69 years) underwent a daily serum monitoring during 14 days for the following parameters: interleukin-6 (IL-6) and TNF (both measured by an ELISA) and CRP (laser nephelometry). Patients were judged to have edematous (AIP) in 6 and necrotizing pancreatitis (NP) in 9 patients.

Results. Significant differences ($p < 0.00001$) could be calculated between AIP and NP for IL-6 in the first 5 days (only the preoperative values were included) after the onset of symptoms of acute pancreatitis (IL-6: median NP 319 pg/ml, range 0–2600; median AIP 0 pg/ml, range 0–44.6). For TNF in both groups only slightly elevated values were found. The efficiency for NP by means of IL-6 (cutoff level: 13.5 pg/ml) and CRP (180 mg/l) were 96 and 86%, respectively.

Conclusion. IL-6 seems to be a very potent parameter to differentiate between AIP and NP comparable with CRP. The high values of IL-6 from the beginning on in NP reflect the leukocyte stimulation in response to the inflammatory process.

145 Adjuvant Radio-Chemotherapy Following Resection of Pancreatic Adenocarcinoma: Preliminary Results

UK Pancreatic Cancer Trials Group (J.P. Neoptolemos)
Department of Surgery, Dudley Road Hospital, Birmingham, UK

Resection for pancreatic cancer is associated with a 5-year actuarial survival rate (YASR) of 5–15%. Adjuvant therapy may be useful but there has only been one study published. The N. American GITSG phase III study reported an improvement following postoperative radiotherapy and 5-FU i.v. weekly in 20 patients with pancreatic cancer (43% 2 YASR vs. 18% in control arm).

The GITSG protocol was repeated in 27 patients with adenocarcinoma of the pancreas ($n = 23$) or periampullary adenocarcinoma ($n = 4$; one with resected solitary liver metastasis). There were 16 men and 11 women with a mean age of 59.4 (range 40–76) years. A standard Kausch-Whipples operation was performed in 11, 13 had a pylorus preserving head resection, 2 had a total and 1 had a distal pancreatectomy. Lymph node metastases were present in 17 cases; the histological grade was $1 = 8$, $2 = 11$ and $3 = 8$. A split course of 40 Gy was given postoperatively over 5 weeks with 5-FU 500 mg/m^2 with each course followed by weekly 5-FU 450 mg/m^2 i.v. The radiotherapy was well tolerated with only one related complication. There was no significant drug toxicity. At a median follow-up of 12 (range 3–36) months the 2 YASR was 30%; 14 patients were dead, 12 alive and 1 lost to follow-up.

These results indicate that postoperative radiochemotherapy is well tolerated. The 2 YASR was intermediate between the treated and control arms of the GITSG study. A larger phase III trial is necessary to establish the efficacy of the GITSG protocol.

146 Relationship between Pancreatic Necrosis Extension and Sepsis in Acute Pancreatitis (AP). Results of a Prospective Study with CT Scan

G. Uomo, M. Laccetti, B. Marcopido, I. Carraturo, G. Manes, L. Pacelli, A. Russo, P.G. Rabitti
Pancreatic Disease Center, Cardarelli Hospital, Napoli, Italy

CT scan with contrast enhancement can identify a high risk of the development of local complications in patients with AP but it cannot reliably predict the infection of the pancreatic necrosis (PN).

In this prospective study we tried to quantify the relationship between the extension of PN evaluated at CT scan and the risk of the sepsis development.

In 32 months we observed 218 cases of AP (96 men, 122 women; mean age 58 years), studied with contrast-enhancement CT within 96 h of onset. The patients were classified in (A) oedematous (163 cases) and (B) necrotizing (55 cases) AP. Group B was divided in (B1) necrosis of the gland $\leq$30% (26 cases), (B2) necrosis up to 50% (14 cases) and (B3) necrosis >50% (15 cases); moreover we identified the patients (n=25) with multiple pancreatic fluid collections (PFC) and those (n=20) with single or absent PFC. In group A there were no pancreatic sepsis and mortality; in group B we found 11 cases of infection of PN; the mortality was 20% (11 cases). Statistical analysis (χ^2 test) showed that the extension of the PN was significantly correlated to the septic evolution (χ^2=12.4; df=2; p=0.002). Instead no correlation was found between the presence and the number of PFC and the infection of the PN.

Our data suggest that the early contrast-enhancement CT scan is a useful prognostic index in foreseeing the septic evolution of the PN in AP.

147 Synthesis and Degradation of Collagen in Pancreatic Fibrogenesis

R. Valderrama, S. Navarro, J.M. López, L. Fernandez-Cruz, A. Giménez, A. Parés, J. Caballería
Gastroenterology Service, Hospital Clínic i Provincial, Barcelona, Spain

Aim. To analyze the role of collagenase in the development of fibrosis in chronic pancreatitis (CP).

Patients. 19 patients with CP and 11 controls (organ donors) with normal pancreatic histology were included in the study.

Methods. Pancreatic tissue was obtained from all subjects to measure: (1) area of fibtrosis (histomorphometric method), (2) prolylhydroxylase activity (POH-ase), which reflects the intracellular synthesis of collagen (Hutton's method), and (3) collagenase activity, which reflects the degradation of collagen (collagenase assay system ^{3}H).

A significant correlation was observed between percent fibrosis evaluated histomorphometrically and POH-ase in all patients (r=0.721; p<0.001), and between POH-ase and collagenase activity in controls (r=–0.702; p=0.024) but not in CP.

Conclusion. Pancreatic tissue in CP has an increased fibrogenetic activity and an impaired collagen degradation capacity. These findings might explain the excessive development of fibrosis in CP.

Study supported by grant from FISss 89/646 (Ministerio de Sanidad y Consumo, Spain).

Results. (Table to Abstract 147)

	Controls (n=11)	CP (n=19)	p
Fibrosis area/total area (%)	4.6±1.8	70.6±20.2	<0.001
POH-ase (cpm/mg protein)	405 ±151	775 ±258	<0.001
Collagenase activity (% digestion/mg tissue)	18.0±3.9	8.7±3.5	<0.001

148 Serum Levels of Selenium (S-Se) in Patients with Chronic Pancreatitis and Varying Degrees of Exocrine Insufficiency

B. Vaona, G. Talamini, A.M. Stanzial[a], *P. Bovo, M. Casaril*[a], *C. Menini*[a], *L. Rigo, L. Frulloni, V. Di Francesco, R. Corrocher*[a], *G. Cavallini*
Clinica Medica and [a]Patologia Medica, University of Verona, Italy

The aim of the study was to assess S-Se values in patients with chronic alcoholic pancreatitis (CAP) with varying degrees of exocrine functional insufficiency. 38 patients (34 M, 4 F; mean age: 48.4±11 years) with well documented CAP were studied; the patients, asymptomatic for at least 3 months, not operated on, with a median follow-up of 10 years, in therapy with enzymes when necessary, were submitted to functional study with S-Cae and then divided into 3 groups: 1) normal (12 patients), 2) moderate insufficiency (11 patients), 3) severe insufficiency (15 patients) (lipase output >41000, 41000–6000, <6000 IU/30 min, respectively). S-Se values in CAP were compared with those of 48 controls (C) (43 M, 5 F; mean age 46.5±8 years). S-Se was determined by atomic absorption spectroscopy (GBC mod. 903). A statistically significant difference (p ranging from 0.007 to 0.00001) was found for the CAP group as a whole (75.8±23.3 ng/ml) and for the groups 2 and 3 vs. C (95.5±13.3 ng/ml). No statistically significant difference emerged between the respective CAP groups with different degrees of insufficiency. No correlation could be established between S-Se and duodenal lipase. In conclusion, S-Se levels appear to be lower in CAP, but do not correlate with pancreatic insufficiency. The intestinal absorption of S-Se may be unrelated to exocrine pancreatic function or may be influenced by enzyme replacement treatment.

149 Alterations of the Exocrine Pancreas in End-Stage Renal Disease

M. Ventrucci, S. Li Bassi, A. DiGrazia, C. Campieri, S. Stefoni, R. Bergami, T. Conci
Departments of Gastroenterology and Nephrology, University of Bologna, Italy

We studied pancreatic function and morphology in 28 patients with end-stage renal disease undergoing regular hemodialysis (twice a week in 4, 3 times a week in 24; mean duration 5 years, range 4 months to 18 years), in order to look for evidence of a specific uremic exocrine pancreopathy. Seventy-nine healthy subjects matched for age and sex served as controls. Studies included: abdominal ultrasonography (US); serum amylase, isoamylases and lipase assays; plus fecal chymotrypsin measurement. None of the patients had clinical evidence of pancreatic disease, but one had had previous attacks of pancreatitis and another developed acute pancreatitis one month after the study. The mean AP diameter of the pancreas body (US) did not differ significantly between patients and controls; however, in 9 patients, size values exceeded those of the control range (4 higher, 5 lower). One patient had a small cyst in the pancreas head, another a slightly dilated pancreatic duct. Mean serum enzyme levels were significantly higher in patients than in controls (amylase, pancreatic isoamylase, and lipase $p<0.001$; salivary isoamylase $p<0.05$). Serum amylase was elevated in 12/28 patients; pancreatic isoamylase in 15/28 (more than twice the upper normal limit in 3); and lipase in 11/28 (more than twice the upper normal limit in 4 patients). Mean fecal chymotrypsin was significantly lower ($p<0.001$) in patients than in controls: abnormally low values were found in 7/28. No correlation was found between the duration of dialysis and pancreatic size (US), serum enzyme levels or fecal chymotrypsin. The frequency of dialysis (2 vs. 3 times/week) did not seem to significantly influence test outcomes. The results indicate that serum pancreatic enzyme elevations are seen rather frequently in uremic patients undergoing dialysis, but that these elevations, as well as less common morphological alterations of the pancreas, are usually mild and nonspecific. The most important finding of this study is the significant decrease in fecal chymotrypsin, which clearly demonstrates an impairment of exocrine pancreatic function in end-stage renal disease.

150 Diabetes Mellitus (DM) after Surgery for Chronic Pancreatitis. Internal Derivations vs. Left Pancreatectomy: A Retrospective Review on 341 Operated Patients

S. Vesentini[a], G. Talamini[b], C. Bassi[a], B. Vaona[b], G. Cavallini[b], P. Bovo[b], F. Nifosì[a], R. Girelli[a], M. Falconi[a], P. Pederzoli[a]
[a]Surgical and [b]Medical Department of the University of Verona, Italy

One of the most trusted beliefs about surgical treatment of chronic pancreatitis is that internal derivations are superior to left pancreatectomy insofar the former are less prone to the late development of DM.

In recent years we were able to collect complete follow-up data on 341 patients operated for chronic pancreatitis. Among them 268 individuals had no biochemical signs of DM (basal glucose and OGTT within normal ranges). Of these 268 non-diabetic patients, 229 (195 men, 34 women, mean age 41.8 years, median preoperative observation 4.2 years) were submitted to an internal derivative procedure (pancreo- or cystojejunostomy or both) and the remaining 39 (31 men, 8 women, mean age 39.7 years, median preoperative observation 4.1 years) were submitted to left pancreatectomy.

After a median postoperative interval of 10 and 11 years, respectively, the two groups were evaluated and compared for the postoperative incidence of DM. Nearly 50% of the patients were overtly diabetics but the Kaplan-Meier test failed to show any significative difference between the two groups.

Should these data be confirmed, the fear for the late development of postoperative DM is no longer a major drawback to the performance of left pancreatectomies in the setting of chronic pancreatitis.

151 Up-Regulation of Somatostatin Receptors by EGF and Gastrin in AR42J Cells

C. Vidal[b], J.P. Esteve[a], F. Lluis[b], N. Vaysse[a], C. Susini[a]
[a]Inserm U 151, CHU Rangueil, Toulouse, France; [b]Laboratorio Investigación Gastrointestinal, Hospital de Sant Pau, Barcelona, Spain

Heterologous regulation of membrane receptors is an important phenomenon in the physiological control of peptide action. We and others have shown that epidermal growth factor (EGF) and gastrin are mitogenic peptides for rat tumor acinar AR42J cells and that somatostatin analog, SMS, suppressed the proliferative effect of EGF [Endocrinology 1989; Cancer Res 1990]. This study was designed to determine whether the growth factors, EGF and gastrin, can modulate somatostatin receptors in AR42J cells and to investigate the mechanism involved.

When cells were grown in serum free medium, treatment of cells with 10 nM EGF or gastrin for 48 h induced an increase of somatostatin binding by 28 and 50%, respectively. ^{125}I-Tyr$_3$ SMS was used as tracer for binding assay. Scatchard analysis showed an increase in somatostatin receptor number (EGF: +21±2%; gastrin: +82±6%) with no change in receptor affinity. The effect of EGF and gastrin was dose-dependent (EC_{50} EGF: 0.2 nM; gastrin: 0.3 nM) and was blocked by pretreatment of cells with 0.01 mM cycloheximide for 24 h. Forskolin or VIP had no effect on somatostatin binding. The phorbol ester TPA did not affect control or gastrin-stimulated binding but reduced EGF-stimulated binding by 70%. Vanadate, a tyrosine phosphatase inhibitor, stimulated AR42J cell growth and increased somatostatin binding in a dose-dependent manner.

In conclusion, EGF and gastrin up-regulate somatostatin receptors by a process involving protein synthesis and phosphorylation of proteins on tyrosine residues. Upregulation of somatostatin receptors may be a physiologic event controlling EGF and gastrin-mediated mitogenic signalling.

152 Source of Circulating Phospholipase A_2 in Necrotizing Pancreatitis

M. Vogeser, M. Büchler, W. Uhl, H.G. Beger
Department of Surgery, University of Ulm, FRG

Free phospholipase A_2 (PLA_2) activities seem to play an essential role in the development of systemic complications in necrotizing acute pancreatitis. In our study serum PLA_2 catalytic activity was compared in patients with necrotizing pancreatitis, diffuse peritonitis and in a control group.

Patients and Methods. Thirty-five patients with acute necrotizing pancreatitis (NP), 21 patients with diffuse peritonitis (PE) and 16 patients after major abdominal surgery as control group (C) were included. PLA_2 was measured by an enzymatic method [J Clin Chem Clin Biochem 1987;25:505].

Results. In diffuse PE as well as in NP we found high PLA_2 activities. C did not show a rise in PLA_2 activity.

PLA_2 catalytic activity (U/l) (medians, quartiles)

	Day 1	Day 2	Day 3
NP	29 (15–42)	20 (15–32)	19 (16–25)
PE	23 (14–46)	18 (9–23)	16 (8–24)
C	5 (0–8)	8 (1–12)	3 (1–9)

Conclusion. Our results support the idea that the main source of PLA_2 in NP may *not* be the pancreas [Clin Chem 1988;34:1052]. Possible sources are leukocytes and the liver (acute-phase protein). There is no elevation of circulating PLA_2 activity during the common postaggression syndrome (control group).

153 Microvascular Ischemia/Reperfusion Injury in the Pancreas of Rats

B. Vollmar, T. Hoffmann, M.D. Menger, K. Messmer
Institute for Surgical Research, University of Munich, FRG

Aim. Based on the concept of 'ischemic injury' as factor initiating acute pancreatitis, we have quantitatively analyzed pancreatic microcirculation in rats during ischemia/reperfusion.

Methods. Under chloralose anesthesia and mechanical ventilation Sprague-Dawley rats were laparotomized. The pancreas was exteriorized for in vivo fluorescence microscopy. Capillary perfusion was assessed by i.v. injection of FITC-hydroxyethylstarch (15 mg/kg; MW 200/0.6), leukocyte-endothelium interaction was documented after in vivo staining of white blood cells by rhodamine-G (1 μmol/kg i.v.). Following baseline recordings ischemia was induced by application of microvascular clamps to the superior pancreaticoduodenal and splenic arteries for 60 min followed by 120 min of reperfusion. Quantitative assessment of capillary density, heterogeneity of capillary perfusion and leukocyte-endothelium interaction was performed off-line by frame to frame analysis, using stereological techniques. Values are given as means±SEM (n=6).

Results. Macrohemodynamics remained unchanged throughout the experimental time. Postischemic reperfusion of the pancreas was characterized by a significant ($p<0.05$) reduction of functional capillary density from 340±12 cm^{-1} during baseline to 270±21 cm^{-1} and 230±64 cm^{-1} after 30 and 120 min of reperfusion. Heterogeneity of capillary perfusion increased by 22% (n.s.) during early reperfusion. Impairment of capillary perfusion was accompanied by a significant ($p<0.01$) increase of permanently adherent leukocytes in postcapillary venules from 21±21 mm^{-2} endothelial surface during baseline to 1,558±333 mm^{-2} and 2,180±368 mm^{-2} after 30 and 120 min of reperfusion, respectively.

Conclusion. Microvascular perfusion failure, heterogeneity of capillary perfusion and leukocyte-endothelium interaction are the microcirculatory characteristics of postischemic reperfusion injury of the pancreas and may trigger the development of acute pancreatitis. Therefore, preservation of pancreatic microcirculation should be the primary goal of therapy for pancreatitis.

154 Effect of Hes and Hyper-Hes on Pancreatic Microcirculation during Hemorrhagic Shock and Resuscitation

B. Vollmar, G. Preissler, M.D. Menger, K. Messmer
Institute for Surgical Research, University of Munich, FRG

Aim. Deterioration of pancreatic blood perfusion during low flow shock state may result in microcirculatory disturbances and focal ischemia which are thought to be causative factors for initiation and/or progression of pancreatitis. Therefore, major intention of resuscitation should focus on preservation of pancreatic microcirculation. This study analyzes in vivo the effects of isovolemic Hes and small-volume Hyper-Hes resuscitation on pancreatic microcirculation after hemorrhagic shock.

Methods. Under chloralose anesthesia and mechanical ventilation Sprague-Dawley rats were laparotomized. The pancreas was exteriorized for in vivo fluorescence microscopy. Capillary perfusion was assessed by i.v. injection of FITC-hydroxyethylstarch (15 mg/kg; MW 200/0.6). Following steady-state conditions, shock was induced by withdrawal of blood to MAP of 40 mmHg and maintained for 1 h. Small-volume resuscitation was performed by replacement of 10% of the withdrawn blood by Hyper-Hes (10% hydroxyethylstarch, 200/0.6, 2,465 mosm/l), isovolemic resuscitation by isotonic hydroxyethylstarch (10%, 200/0.6, 308 mosm/l). Macro- and microhemodynamic parameters were analyzed during baseline, at 1 h shock, 20 min and 1 h after resuscitation, respectively.

Results. MAP regained 57% (Hyper-Hes) and 69% (Hes) of baseline values after 20 min of resuscitation and decreased to 42% and 60% of baseline after 1 h. In both groups, 1 h low flow shock state was characterized by a marked reduction of functional capillary density of the pancreas to 68% of baseline values. After 20 min of resuscitation with 2 ml/kg Hyper-Hes or 20 ml/kg Hes, almost complete restoration of pancreatic microvascular perfusion was observed (88 and 85%, respectively). After 1 h resuscitation functional capillary density was 77 and 83% in Hyper-Hes and Hes-treated animals.

Conclusion. In low-flow shock state pancreatic microcirculation is characterized by a significant decrease of functional capillary density. Despite replacement of only 10% of actual volume loss, small volume resuscitation with Hyper-Hes proved to be as effective as isovolemic resuscitation with Hes in *early* restoration of pancreatic capillary perfusion.

155 Cerulein Hyperstimulation in vitro Induces Enhanced Expression of a 70-kD Heat Shock Protein in Isolated Pancreatic Lobuli

C. Weber, T. Gress, F. Müller-Pillasch, H. Weidenbach, G. Adler
Abteilung Innere Medizin I, University of Ulm, FRG

Heat shock proteins (HSP) have been shown to be synthesized by mammalian cells as a reaction to a variety of exogeneous stressors. HSP70 for example plays a major role in facilitating the folding and transport of intracellular proteins under conditions leading to cell injury [Rothman JE, Cell 1989;59:591–601]. To examine the involvement of heat shock proteins during cerulein hyperstimulation of pancreatic tissue we examined the induction of HSP70 and ubiquitin transcription in rat pancreatic lobuli in vitro. We chose this in vitro model to eliminate alterations due to inflammatory cells. Freshly isolated rat pancreatic lobuli were incubated with RPMI medium at 37 °C. Cerulein was added at concentrations of 100 n*M*. Unstimulated lobuli were heat shocked at 42 °C for 1 h and served as a positive control. After 2 h of incubation RNA was extracted and transcript levels of HSP70 and ubiquitin were assessed by Northern blot analysis. In addition we examined the expression of HSP70 by immunofluorescence techniques using a monoclonal antibody.

Results. Both immunofluorescence and Northern blot analysis revealed an increased turnover of HSP70 in cerulein treated pancreatic lobuli as compared to unstimulated controls. Heat shocked tissue showed the highest expression of HSP70. No significant differences of ubiquitin expression were observed in all 3 experimental approaches. We conclude that in vitro secretagogue stimulation induces an increased synthesis of HSP70 in rat pancreatic tissue. Although transcriptional levels in cerulein treated lobuli were much lower as compared to heat shocked tissue, HSP70 might play an important protective role during cellular injury of pancreatic tissue caused by cerulein hyperstimulation.

••••••••••••••••••••

156 Time-Dependent Expression of Extracellular Matrix Components in the Human Pancreas during Fetal Development

A. Zschäbitz[a], *M. Neurath*[a], *H. Printz*[b], *W. Schmidt*[c], *E. Stofft*[a]
Departments of Anatomy, Universities of [a]Mainz and [c]Leipzig; [b]Department of Surgery, University of Marburg, FRG

The extracellular matrix is part of a complex recognition system that governs the social behavior of pancreatic cells. In the present study, we have attempted to map different matrix components during fetal pancreas development and to compare the results with adult pancreatic tissue.

Methods. Samples from 18 human fetuses ranging from 3 to 8 months (mostly late abortion) and adult pancreatic tissue obtained from organ donors (n = 13) were examined by light, transmission, and scanning electron microscopy. The expression of endogenous and exogenous lectins and several structural glycoproteins (laminin, fibronectin, tenascin, collagen types III, IV, VI) was determined by immunohistochemistry using isothiocynat or streptavidine-conjugated secondary antibodies.

Results. The proof of terminal Gal-β(1–3)GalNAc molecules in the pericellular matrix was confined to early fetal stages. The content of sialic acid residues and non-sulfated acid proteoglycans in the periductal matrix increased during fetal development. Type VI collagen was found from the early stages of development. Type III collagen was detected at the end of the 3rd month. Its binding intensity was augmented during organogenesis. At the 4th month type IV collagen became present as a thin layer underlaying vessels and ducts. At the same developmental stage receptors for fucose binding LTA appeared on the endothelial surface. In adult pancreas the binding affinity of UEA I but not of LTA correlated with the type IV collagen arrangement. Laminin binding patterns remained unchanged. Tenascin decreased significantly during early fetal development. In contrast, the staining intensity for fibronectin was significantly reduced not before the 7th fetal month. In adult specimens, the matrix was only weakly marked using mAbs against both adhesion molecules. In fetal tissue, WGA revealed a higher affinity for the extracellular ground substance than in adult pancreas. Endogenous lectins binding terminal mannose groups decreased during fetal development. Endogenous matrix receptors with an affinity for sialic acid molecules were increased at later stages of development. There was no difference in the binding patterns between fetal pancreas and adult specimens using the lectins Con A, PHA E, and RCA and the neoglycoproteins N-acetylgalactosamine, xylose and heparan.

Conclusions. The results demonstrate a characteristic topographical expression of extracellular matrix components during the development of the human pancreas. Fetal pancreas also contains endogenous lectins. The regulated expression of glycoconjugates suggests their important regulatory function.

Author Index

Numbers refer to Abstract No.

Digestion 1992;52:138–144

I. Ihse
C.W. Imrie
P.G. Lankisch
Å. Andrén-Sandberg
L. Gullo
R. Laugier

Department of Surgery,
University Hospital, Lund,
Sweden

Diagnosis of Exocrine Pancreatic Diseases

Working Party Report to the European Pancreatic Club, Lund 1991

Key Words
Acute pancreatitis
Chronic pancreatitis
Diagnosis
Pancreatic cancer

The following suggestions and recommendations for the diagnosis of acute (AP) and chronic pancreatitis (CP) as well as for pancreatic carcinoma (PC) were developed by a Working Party to the European Pancreatic Club. Senior members of the club served as an editorial board, and the paper was submitted to them for review and then for discussion at the meeting of the European Pancreatic Club in Lund, Sweden. The recommendations were revised according to the changes suggested by the reviewers and discussion participants. The aim of the Working Party was to develop a standardized basis for multicenter clinical studies as well as to facilitate the comparison of center-to-center and country-to-country diagnoses. It was also meant to aid the nonspecialized colleague in the diagnosis of AP and CP as well as PC.

Acute Pancreatitis

The clinical presentation, biochemical and morphological changes in AP vary according to the severity of the disease. Approximately 20–25% of patients have severe disease, while the condition is mild in 75–80%. There is a range of presentation from fulminant severe AP, which is immediately life-threatening to the mildest form, which clinically resolves in 3 or 4 days. Patients at each end of the spectrum of disease are relatively easy to categorize, while a large group in the center is difficult to

Received:
February 20, 1992
Accepted:
June 25, 1992

I. Ihse
Department of Surgery
University Hospital
S-221 85 Lund (Sweden)

0012–2823/92/
0522–0138 $ 2.75/0

grade at the initial presentation. Thus, objective grading systems are not only justifiable, but help greatly in allocating the severe group to receive high dependency on intensive care treatment with appropriate allocation of expensive investigations, such as angiogram enhanced CT scanning. Objective grading systems forewarn the clinician of a high probability of major complications of AP, as well as a higher mortality rate in those patients graded to have severe disease.

Symptoms and Signs

Upper abdominal pain which is severe and of sudden onset characterizes 90% of patients. This pain may be in the epigastrium, either or both upper quadrants of the abdomen, and it may radiate into the back. Occasionally, it encircles the whole of the upper abdomen. Less severe pain or discomfort of gradual onset characterizes the remaining small number of patients. Pain which presents high in the left upper quadrant of the abdomen may be difficult to differentiate from severe angina or myocardial infarction. The pain of AP tends to be unrelenting until strong analgesics are given by a parenteral route.

Vomiting is a characteristic of over 85% of patients, and this may be so severe that it causes even more trouble to the patient than pain.

On examination: abdominal guarding, rigidity, rebound tenderness and the loss of bowel sounds are characteristic. However, in the very early phase after onset of AP, some of the signs may be absent.

In the most severely ill, pallor, sweating, tachycardia with hypotension and a fast respiratory rate are often found. Hypovolemia is a major feature with vomiting and loss of intravascular albumin as the main elements.

It should be noted that lip cyanosis is not a characteristic feature, unless a preexisting cardiorespiratory disease exists.

Laboratory Investigations

Characteristically, gross elevations in pancreatic enzymes are to be found. In practice, total serum amylase and serum lipase are the most valuable tests. Ideally, both should be carried out, but if only one is to be chosen, then the lipase measurement is more specific for the diagnosis of AP. If only the serum amylase is measured, then a moderate degree of elevation (between one and three times the upper limit of normal) should encourage the clinician to obtain a urinary amylase estimation and a repeat serum amylase estimation in a few hours. In selected cases, where hyperlipemia is present, or there is a suspicion of macroamylasemia, then a urinary amylase estimation may also be very valuable. Estimations of pancreatic isoamylase may be helpful, but are not generally available.

Imaging Procedures

The differential diagnosis of the emergency presentation of AP is with perforated peptic ulcer, mesenteric ischemia or infarction, acute cholecystitis, dissecting aortic aneurysm and ectopic pregnancy. All of these (with the exception of acute cholecystitis) can be associated with significant elevations in serum amylase. An early plain X-ray of the abdomen will show free gas in 65% of those with perforated peptic ulcer, while early ultrasonography will usually detect stones in the biliary tree. An initially negative ultrasound scan for stones should be repeated in 2–3 days, as it has been shown that a further significant percentage of stones is detected at that time. Ultrasound may also show bile duct dilatation, biliary duct stones and abnormal pancreatic morphology. The plain X-ray should differentiate AP from small bowel obstruction.

Objective Grading of Severity of AP

Peritoneal aspiration is a useful method, particularly in those with alcohol-induced AP, those patients with severe disease having dark-coloured fluid free from bacteria. This system is poor in the gallstone group of patients. The original Ranson system is also poor for the gallstone group of patients, and the double system, which is subsequently described in 1981, is somewhat impractical. The Glasgow system of eight prognostic factors has been shown to have an approximately similar specificity for alcohol- and gallstone-induced AP between 65 and 70% accuracy. This and the Ranson system are still of value, but will soon be replaced by new methods of assessment. The presence of three or more prognostic factors in each system indicates severe AP. The greater the number of prognostic factors the worse the prognosis. APACHE-II grading is also rather cumbersome, and different weighting for the age and previous health of the patient is necessary. Someone skilled in using the system can obtain a good separation between the patients with severe and mild disease, scores of greater than eight typify severe AP.

C-reactive protein is a very valuable test which is cheap to perform and readily available in all hospitals. It separates the group of severe and mild disease accurately from 36 h after the onset. Unfortunately, the separation between the two groups is not satisfactory on admission soon after the onset of pain. Levels greater than 150 mg/l typify severe AP, and many patients in the severe group have levels over 250 mg/l. Leukocyte (PMN) elastase shows great promise to be the most valuable marker of severity of disease. It has recently become available as a test from a single manufacturer. It has the major advantage of being markedly elevated in severe disease at the time of admission to hospital, while only moderate elevations are found in mild AP. At the present time, it has only been employed in two pilot studies (in Spain and Germany) and requires fuller assessment before total endorsement.

The Working Party's opinion was that all clinicians should employ one or more objective grading system and that patients graded to have severe AP should be moved to centers of expertise in the management of the complicated disease where angiogram enhanced CT scans and endoscopic retrograde cholangiography (ERCP) facilities are available. If the above guidelines are followed, then laparotomy diagnosis should very rarely be necessary.

CT scanning of the pancreas should always be accompanied by an intravenous bolus injection of contrast to provide an angiogram enhanced picture to assess the perfusion of the pancreas. In the late presentation, CT scanning can confirm the diagnosis of AP, if pancreatic swelling and the accumulation of fluid in the peripancreatic and perirenal spaces are present. The additional importance of CT scanning in severe AP is the delineation of poorly perfused or nonperfused areas of the pancreas, the early detection of peripancreatic gas collection and the facility for guided needle aspiration in patients being considered for surgical intervention. Objective grading of severity based on CT scan scoring systems are available, but it is not recommended for general use on the basis of expense.

The availability of expert ERCP facilities in severe AP where a gallstone etiology is probable or proven is considered highly advantageous, not only to confirm the gallstone etiology but, to remove any stone lodging in the common bile duct.

Chronic Pancreatitis

Symptoms and Signs

The classical symptoms of CP are pain and exocrine and endocrine pancreatic insufficiency. Exocrine pancreatic insufficiency of different degrees is always present, whereas pain by definition is absent in painless CP

(5–8% of all cases of CP). The majority of patients with painless CP have endocrine pancreatic insufficiency, mostly insulin-requiring diabetes mellitus.

Pain is usually intense, constant, lasting at least 10 h and localized in the epigastric and/or periumbilical area, radiating to the back. Other locations are possible. Pain may be combined with nausea and vomiting; relapses at unpredictable intervals and of unpredictable intensity may occur. In a number of patients, particular in the presence of local complications, such as pancreatic duct abnormalities, pseudocysts and biliary obstruction, pain occurs approximately 0.5 h after a large meal with or without alcohol. Eating usually increases an already existing pain. Pain can be partly relieved by bending over or pulling up the knees. The first pain attack in CP may be an acute one. If a patient suffers pain and admits consuming larger quantities of alcohol on a regular basis, the likelihood of CP as the underlying diagnosis is high. Pancreatitis-like pain in patients over 65 years of age is almost never due to CP, but indicative of PC.

Both weight loss and fatty stools are frequent symptoms in the late stages of the disease. Weight loss may be an early symptom during pain attacks when the patient does not eat or drink.

Diabetes mellitus may occur in the late stages of the disease, but is usually one of the leading symptoms in painless CP.

Jaundice is a possible transient symptom due to an acute inflammatory swelling of the head of the common bile duct.

There are, unfortunately, no specific clinical signs of CP. During an acute attack, an inflammatory painfull mass in the upper abdomen may be palpable.

Laboratory Investigations

There are no specific laboratory tests. Serum enzymes (amylase, lipase) may be elevated during acute attacks, especially during the early phase of the disease, but normal test results do not exclude CP. If specific pancreatic enzymes in serum (pancreatic isoamylase, lipase, trypsin) are low, exocrine pancreatic insufficiency is very likely. However, due to the low sensitivity of these tests, they are not useful for screening.

Imaging Procedures

When pancreatic calcifications are present in abdominal X-rays, the diagnosis is firmly established. Ultrasonography is mandatory in patients with pain suspected to be of pancreatic origin as well as in those with known CP in order to diagnose possible complications of the disease (pancreatic calcifications, ductal changes, cysts, abscesses). The investigator's experience is of utmost importance for the test result. In the early stage of the disease, the ultrasonography of the gland is usually normal. Computed tomography is only indicated in patients in whom repeated ultrasound examinations fail to fully visualize the gland or in whom PC on the basis of CP is suspected. ERCP is an invasive examination which is not absolutely necessary for diagnosis of CP, but may act as a guide if surgery is planned. Furthermore, ERCP may exclude common bile duct disease which may lead to chronic pancreatitis-like pain. Prior to ERCP, a gastroscopy should be performed to exclude diseases of the upper gastrointestinal tract which may also simulate the presence of CP. In order to classify the results of imaging procedures, the Cambridge classification of pancreatography [1], of ultrasonography or of computed tomography [2] may be used.

Pancreatic Function Tests

Pancreatic function tests may be divided into direct and indirect tests. The secretin-pancreozymin test with its several modifications is most sensitive, but time-consuming, invasive, expensive, and thus not everywhere available. The same applies for the Lundh test, which is considered to be less sensitive than the secretin-pancreozymin test. However, this has never been proven beyond doubt. Usually, the clinician will refer to indirect tests (fecal chymotrypsin estimation, BT-PABA test, pancreolauryl test), the results of which are usually abnormal in moderate and severe exocrine pancreatic insufficiency. Thus, a normal indirect test result excludes moderate/severe exocrine pancreatic insufficiency. However, in slight functional impairment – generally in the early phase of CP – the indirect tests may be falsely normal. In the early phase, a direct test should be chosen because of its higher sensitivity.

Recommended Diagnostic Steps in Suspected CP

The early stage of the disease can be characterized as follows: clinical history: suspected CP; clinical examination: usually negative; laboratory findings: serum enzymes elevated during acute attacks, otherwise not helpful; plain abdominal X-ray: usually normal; ultrasound: usually normal; pancreatic function tests: direct tests may be the first abnormal tests, indirect tests may be normal in slight exocrine pancreatic insufficiency, and ERCP: if indirect pancreatic function tests are normal and direct pancreatic function tests not available, but pain is still the leading symptom, ERCP should be performed, but may be normal.

It should be noted that even if pancreatic function tests and morphological examination are normal, CP is not ruled out. Renewed investigation is recommended every 6 months in accordance with the evolution of the disease.

The features of the advanced stage of the disease with and without complications are as follows: clinical history: usually positive; clinical examination: usually no signs, but weight loss and jaundice may be present; a palpable mass in the upper abdomen and ascites are rarely seen; laboratory findings: seldomly helpful, serum enzymes usually normal, but occasionally subnormal; plain abdominal X-ray: calcifications may be present; ultrasonography: *essential,* usually abnormal; pancreatic function tests: direct and indirect tests abnormal, and ERCP: usually abnormal, preoperative examination is the main indication.

Endoscopic ultrasonography, ultrasound or CT-guided percutaneous puncture or NMR examinations are not widely available, require highly specialized investigators, and their role is not yet defined.

Pancreatic Cancer

If the patient is consulting the doctor because of symptoms, the diagnosis PC is nearly always easy to make. However, in spite of the introduction of new and sophisticated diagnostic tools, an early diagnosis is still rare, as all symptoms and signs present in the late course of the development of the disease. Furthermore, there is no test available which is specific and sensitive enough, for screening purposes, and there are no defined risk groups.

Symptoms and Signs

Carcinoma of the head of the pancreas, accounts for 70% and the classic presentation is with painless jaundice, although a degree of discomfort may be present in as many as 50% of patients. When jaundice appears most patients have already had a history of weight loss, which sometimes is extensive (10–35 kg). Occasionally, patients present with AP or hyperamylasemia associated with discomfort.

Carcinoma of the body and/or tail of the pancreas is classically associated with persistent discomfort or pain occurring in the upper lumbar area posteriorly. Anorexia, weight loss and mood changes may also be prominent.

Thrombophlebitis migrans can be associated with any carcinoma of the pancreas, but is not specific to this type of carcinoma. A loss in desire for normal foods or for tobacco (for those who use it) may be an early sign of malignant disease, as can the debut of diabetes mellitus in an individual of normal body weight and without heredity predisposition for diabetes mellitus.

Laboratory Investigations

In patients with painless obstructive jaundice, the normal response to compression of the bile duct is found with high levels of bilirubin, alkaline phosphatase, γ-glutamyltransferase being typical. In a minority high levels of pancreatic enzymes may result from the inflammatory reaction around the tumor or from obstruction of the main pancreatic duct by the tumor, but may also indicate the presence of an acinar cell carcinoma.

Serological markers, such as CA 19-9, CA-50 an CEA, are often elevated in the majority of patients with PC once the tumor exceeds 2 cm in diameter. They are not of value in the early diagnosis of small tumors as the degree of abnormality is proportional to the size of the primary tumor.

Imaging Tests

The best initial investigation is ultrasonography, if done by an experienced examiner, and ERCP ($\pm$cytology or biopsy) is worthwhile, especially in jaundiced patients. Angiogram enhanced CT scanning is complementary to ultrasonography and may provide accurate staging information with regard to the extent of the carcinoma, which will influence therapeutic approaches. Selective angiography does not confer any additional benefit as investigational approaches and is, at the present time, not recommended for routine use. Laparoscopy assessment of the presence or abscence of liver or peritoneal metastases imply an advantage in staging carcinoma of the pancreas prior to a decision on the wisdom of surgical resection. MRI may be a valuable investigational tool in the future. Recent studies suggest that endoscopic ultrasonography detects even small tumors of the pancreas. The equipment and the experience is, however, still restricted to a small number of centers.

Screening Tests

The question of screening 'at risk' patients by means of ultrasonography, CT scan and/or ERCP was discusssed with specific reference to adult onset diabetes in fit patients over 50 years. The unanimous opinion was that such a screening procedure currently cannot be recommended, because of the low yield related to the expense of the screening and the additional anxiety created in those screened.

Cytological or Histological Diagnosis

Definitive diagnosis by means of (a) endoscopical biopsy, (b) ERCP cytology, (c) percutaneous fine needle aspiration cytology or (d) percutaneous biopsy specimens was considered highly desirable in patients who had been graded unsuitable for surgery on the basis of earlier investigations, e.g. if an ultrasonography had demonstrated a primary lesion in the head of the pancreas with liver metastases. It was not considered necessary, prior to resectional surgery, for cytological or histological evidence of malignant disease to be obtained. Indeed considerable concern was expressed at overenthusiastic pursuit of percutaneous methods with possible spillage of tumor cells into the peritoneal cavity, and the difficulty of accurately directing the needle into a small primary tumor. Thus, a negative examination does not preclude the presence of carcinoma and therefore does not contraindicate resectional surgery.

Summary

The Working Party was of the opinion that certain investigations were of little or no clinical value in PC and, outside research projects, there was no place for the routine use of serological markers of disease or selective angiography. The emphasis of this brief report is therefore on a streamlining of investigation with increased expertise in ultrasonography diagnosis, as well as ERCP assessment. More wide-spread availability of angiogram enhanced CT scanning is desirable. In the minority group (approximately 15 per cent) considered suitable for surgical resection preoperative cytology should be avoided.

References

1 Axon ATR, Classen M, Cotton PB, Cremer M, Freeny PC, Lees WR: Pancreatography in chronic pancreatitis: international definitions. Gut 1984;25:1107–1112.
2 Sarner M, Cotton PB: Classification of pancreatitis. Gut 1984;25:756–759.